TABLE OF CONTENTS

I. INTRODUCTION 1
    A. THESIS OBJECTIVES 1
    B. RELATED WORK 1
    C. RESEARCH QUESTIONS 3
    D. THESIS ORGANIZATION 3

II. TECHNOLOGY BACKGROUND 5
    A. TCP 5
        1. TCP Connection 6
        2. Reliable Data Transfer 12
        3. Flow Control 15
        4. Congestion Control 16
            a. Slow Start 17
            b. Fast Retransmit/Fast Recovery 18
            c. Congestion Avoidance 20
        5. TCP Message Format 21
    B. WAN OPTIMIZATION DESCRIPTION 26
        1. Methods of WAN Optimization 28
            a. TCP Acceleration 28
            b. Data De-duplication 34
            c. Application Acceleration 37
            d. Compression 38
            e. Suppression 39
            f. Increasing TCP Active Window Size 40
        2. PERFORMANCE EVALUATION OF WAN OPTIMIZERS 40

III. METHODOLOGY 45
    A. REQUIREMENTS/GOALS 45
        1. Scope 45
    B. DESIGN 46
        1. MySQL Database 49
            a. Remove TCP Retransmissions 50
        2. Two Hosts 50
    C. IMPLEMENTATION 50
        1. PUT ALL IP PACKETS INTO DATABASE 52
            a. Mark packets that are retransmissions ... 54
        2. Write the packets to a separate file for
            sending and receiving hosts 57
        3. orderedReplay.py 57
            a. Test setup for orderedReplay.py 60

IV. EXPERIMENTAL RESULTS 63
    A. TEST SETUP 63
    B. TEST 1 – A SHORT TCP FLOW WITH RETRANSMISSIONS 67

     C.   TEST 2 — 5MB WEB DOWNLOAD 76
     D.   TEST 3 — SAME NETWORK TRACE, DIFFERENT TCP FLOW ...86
     E.   TEST 4 — LARGE FILE DOWNLOAD FROM WEB 96
V.   CONCLUSIONS, RECOMMENDATIONS, AND FUTURE WORK 107
     A.   CONCLUSIONS 107
     B.   RECOMMENDATIONS 108
     C.   FUTURE WORK 109

APPENDIX 111
     A.   SOURCE CODE FOR TCPFLOWPREPPER.PY 111
     B.   SOURCE CODE FOR ORDEREDREPLAY.PY 124
     C.   SOURCE CODE FOR FLOWSTATS.PY 130

LIST OF REFERENCES 135

INITIAL DISTRIBUTION LIST 137

LIST OF FIGURES

Figure 1.   OSI Seven Layer Model. From [4] 6
Figure 2.   TCP Header Format. From[5] 7
Figure 3.   TCP State Transition Diagram. From [6] 9
Figure 4.   Missing Packet Example 15
Figure 5.   TCP Congestion Control Algorithm. After [8] 17
Figure 6.   Fast Retransmit 19
Figure 7.   TCP Congestion Control [8] 20
Figure 8.   TCP Sequence Numbers 22
Figure 9.   TCP Sequence Numbers 24
Figure 10.   Common WAN Optimization Topology 27
Figure 11.   TCP Acceleration 29
Figure 12.   TCP Without Optimization 32
Figure 13.   Data De-duplication 35
Figure 14.   Program Architecture 47
Figure 15.   Sample order.txt 48
Figure 16.   Test Network Configuration 48
Figure 17.   Database Schema[After 18] 53
Figure 18.   Cases for Retransmission 55
Figure 19.   Cases for Retransmission 56
Figure 20.   orderedFlow.py Decision Matrix 59
Figure 21.   Receiving Socket 61
Figure 22.   iptables Rule 67
Figure 23.   Test 1 Target TCP Flow Screen Capture 69
Figure 24.   Screen Capture from Host A for Test 1 70
Figure 25.   Screen Capture from Host B for Test 1 71
Figure 26.   SYN Packet from pcap1.pcap 73
Figure 27.   SYN Packet from orderedReplay.py 74
Figure 28.   TCP Packet with Data from pcap1.pcap 75
Figure 29.   TCP Packet with Data Captured on Host A 76
Figure 30.   First 30 packets of the Source Network Trace
             for Test 2 77
Figure 31.   First 30 packets Network Trace from Host A
             During Test 2 79
Figure 32.   First 30 packets Network Trace from Host B
             During Test 2 80
Figure 33.   Last 30 packets Source Network Trace for Test 2 .81
Figure 34.   Last 30 packets Network Trace from Host A During
             Test 2 82
Figure 35.   Last 30 packets Network Trace from Host B During
             Test 2 83
Figure 36.   Test 2 HTTP Get Message from Source Network
             Trace 84
Figure 37.   Test 2 HTTP Get Message from Host A 85

Figure 38.   Test 2 HTTP Get Message from Host B 86
Figure 39.   First 30 packets from Source Network Trace
             During Test 2 88
Figure 40.   First 30 packets Captured on Host A During Test
             2 89
Figure 41.   First 30 packets Captured on Host B During Test
             2 90
Figure 42.   First 30 packets From Network Trace File 91
Figure 43.   First 30 packets Captured on Host A During Test
             2 92
Figure 44.   First 30 packets Captured on Host B During Test
             2 93
Figure 45.   Test 3 Packet With Data From Source Network
             Trace File 94
Figure 46.   Test 3 Packet With Data From Host A 95
Figure 47.   Test 3 Packet With Data From Host B 96
Figure 48.   First 35 Packets from Source Network Trace File 98
Figure 49.   First 35 Packets Captured at Host A During Test .99
Figure 50.   First 35 Packets Captured at Host B During Test100
Figure 51.   Last 35 Packets from Source Network Trace File 101
Figure 52.   Last 35 Packets Captured from Host A 102
Figure 53.   Last 35 Packets Captured on Host B 103
Figure 54.   TCP Packet with Application Data from Source
             Network Trace File 104
Figure 55.   TCP packet with Application Data Captured on
             Host A 105
Figure 56.   TCP Packet with Application Data Captured on
             Host B 105
Figure 57.   WAN Optimization Testing Architecture 110

LIST OF TABLES

Table 1.   TCP State Transition Diagram. From [6] 10
Table 2.   Network Traces Used in Test 65
Table 3.   Targeted TCP Flow in Each Test 65
Table 4.   Program Performance 66

THIS PAGE INTENTIONALLY LEFT BLANK

LIST OF ACRONYMS AND ABBREVIATIONS

ACK	Acknowledgement
ARP	Address Resolution Protocol
BDP	Bandwidth Delay Product
BER	Bit Error Rate
CWND	Congestion Window
DNS	Domain Name Service
FIN	Finished
FTP	File Transfer Protocol
HTTP	Hyper Text Transport Protocol
IANA	Internet Assigned Numbers Authority
ICANN	Internet Corporation for Assigned Names and Numbers
IEEE	Institute of Electrical and Electronics Engineers
IP	Internet Protocol
LFN	Long Fat Network
LFP	Long Fat Pipe
MAC	Media Access Control
MCB	Marine Corps Base
MCNEL	Marine Corps Network Efficiency Lab
MCTSSA	Marine Corps Tactical System Support Agency
MSS	Maximum Segment Size
NIC	Network Interface Card
OSI	Open Systems Interconnection
RFC	Request for Comments
RST	Reset
RTO	Retransmit Time Out
RTT	Round Trip Timer
RWND	Receive Window
SEQ	Sequence Number

SACK	Selective Acknowledgement
SQL	Structured Query Language
SRTT	Smoothed Round Trip Time
SSTHRESH	Slow Start Threshold
SYN	Synchronization
TCP	Transmission Control Protocol
UDP	User Datagram Protocol
WAN	Wide Area Network

ACKNOWLEDGMENTS

There are many people whom I would like to thank for their help and support in completing my thesis. I would like to thank Capt Pete Young and his staff at MCTSSA, for providing the thesis topic and also for allowing me to visit their facilities to learn about how they conduct their testing. I also would like to thank Mr. Eric Otte and his staff from SPAWAR Systems Center Pacific, for allowing me to visit their facilities to see how they conduct WAN Optimization testing. Others I would like to recognize include Jeff Dean, for his assistance with all things Linux; Ryan Craven, for his assistance with the Python DNET module and his expertise on TCP sockets; Jeff Silverman, for his documentation of the Python module DPKT and for answers to my email questions; and LCDR Rob Yee, LCDR Tony Nichols, LCDR Todd Sehl, Major Josh Bundt, LT Chris Bonine, and LT Bradley Crocker, for providing group support in helping drag each through two years of classes and research.

To my thesis advisors, Geoff Xie and John Gibson: Without your insight, guidance, professionalism, and patience I would not have been able to complete the process. I will do my best to recruit more students for the network track while out in the fleet.

To my wife, Kirsten, and my two kids, Corey and Tierra, thank you for all the love and support over the last two years. There is no way I could have come this far without you guys.

THIS PAGE INTENTIONALLY LEFT BLANK

I. INTRODUCTION

A. THESIS OBJECTIVES

The Marine Corps Tactical System Support Activity (MCTSSA) Marine Corps Network Efficiency Lab (MCNEL) at Camp Pendleton, CA is currently evaluating Wide Area Network (WAN) Optimization and has determined there is a need for a better tool to test WAN Optimization devices using stateful TCP replay. The objective of this thesis is to develop a software prototype that can re-create the two-way TCP flow traffic captured between two hosts in a laboratory setting while retaining the relative ordering of the packets and their TCP payload. The prototype should ensure that tests are repeatable with minimum overhead and that different TCP flows can be replayed in order to test the efficiency of WAN Optimization products. In addition, the software should not replay retransmissions that would not occur with the deployment of WAN Optimization gateways. This thesis lays the groundwork and establishes a framework for the development of a testing capability for measuring the efficiency of different WAN optimization products with respect to an excerpt of real-world production TCP flows.

B. RELATED WORK

Most of the tools developed for stateful replay of TCP traffic are designed for testing network security. Most of these tools replay a capture live trace of TCP traffic from within a network to identify or reproduce specific network events in a test environment.

SocketReplay [1] is a tool that performs loss recovery of packets from a live network capture and replays TCP flows statefully. SocketReplay is designed for large networks and deliberately removes most of the data from the payload in order to reduce the file size of the capture. Our tool uses the entire data payload in each packet in order to re-create the original TCP flow as accurately as possible in order to accurately evaluate some of the WAN optimization techniques. Both tools enable selective replay of TCP flows which allows the user to identify key characteristics about a flow in order to test a network. However, SocketReplay was specifically created for testing network security and detecting network intrusions and viruses, while our tool was specifically created to replay a stateful TCP session in order to test the efficiency of WAN Optimization.

Monkey [2] is a tool that captures live network TCP traces and determines network performance metrics such as delay, bottleneck bandwidth, etc. and then attempts to replay captured Hyper Text Transport Protocol (HTTP) get requests in a lab environment exhibiting similar performance metrics. Monkey tests how changes made to the TCP stack at the server level affect how the server handles large web traffic loads. Monkey-based tests look mainly at replaying short HTTP flows, while our tool is concerned with replaying all types of TCP flows but especially larger data transfers over links with a high bandwidth delay product, that is, performance disadvantaged links.

IXIA Inc. markets a product called BreakingPoint Storm [3] that replays multiple TCP network trace files in a random order. BreakingPoint Storm is a commercially available tool that randomly selects TCP flows from a network and replays them. One conclusion from the MCNEL group's tests of WAN optimization devices at Camp Pendleton is that a random selection of TCP flows caused a high number of TCP errors. For this reason, our tool plays one flow at a time, though it may be extended later to replay multiple flows concurrently.

C.   RESEARCH QUESTIONS

- To summarize, this thesis addresses primarily the following research question: Is it possible to replay a TCP flow in a laboratory setting while retaining the relative ordering of the packets and their TCP payload?

- In addition, this thesis aims to answer the following subsidiary research questions: Is it possible to re-create a TCP flow in a lab setting with the same inter-packet timing as seen in a live TCP packet capture? What are WAN Optimization techniques? What are current methods to measure WAN Optimization results?

D.   THESIS ORGANIZATION

This thesis is organized as follows. Chapter II presents background information on TCP and WAN Optimization. Chapter III discusses the methodology for and implementation for creating the software prototype and testing. Chapter IV analyzes the data that was captured during testing. Chapter V presents conclusions, and makes recommendations for expanding the framework of the software prototype to match the requirements stated by MCTSSA.

THIS PAGE INTENTIONALLY LEFT BLANK

II. TECHNOLOGY BACKGROUND

The purpose of this thesis is to create a software prototype that can re-create the two-way traffic of a captured TCP flow between two hosts in a laboratory setting while retaining the relative ordering of the packets and their TCP payload data. This work lays the groundwork and establishes a framework for the development of a testing capability for measuring the efficiency of different WAN optimization products with respect to an excerpt of real world production TCP flows. This chapter will discuss the basics of the Transmission Control Protocol (TCP), including reliable data transfer, and the mechanisms for flow control and congestion control, and the TCP message format. This chapter will also discuss the need for WAN Optimization, methods of performing WAN Optimization, and methods used to test WAN Optimization products.

A. TCP

TCP was designed to provide reliable end-to-end segment delivery. This means that TCP relies on the end hosts to ensure delivery of all segments and not on the routers between those hosts. TCP is connection-oriented and functions as layer 4 of the Internet Model of internetworking; its functions include those of the transport layer and many of those of the session layer of the Open Systems Interconnection (OSI) Model (see Figure 1).

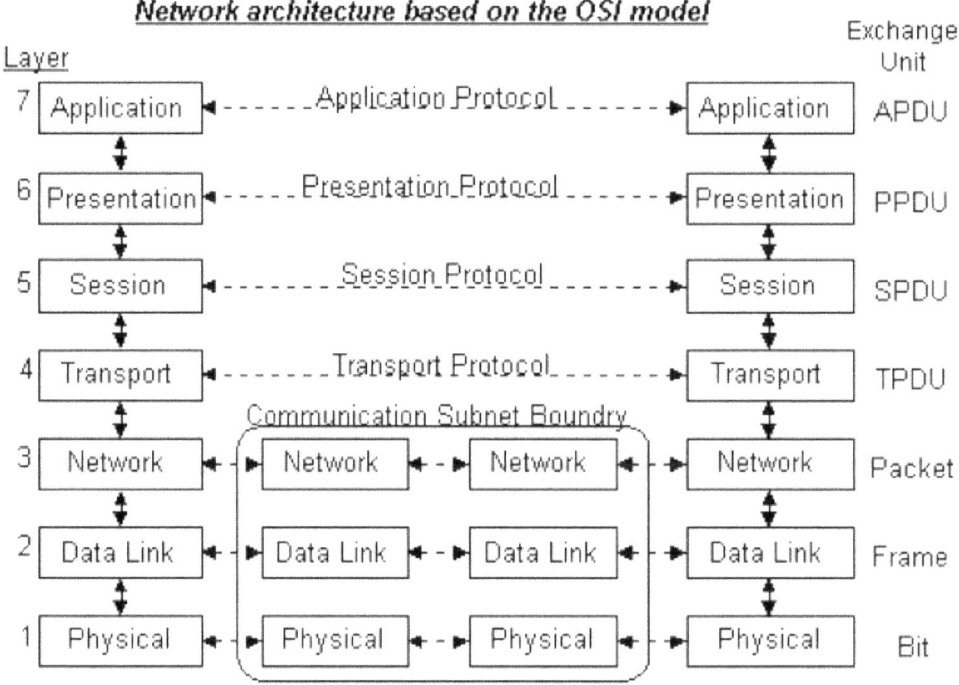

Figure 1.   OSI Seven Layer Model. From [4]

1.   TCP Connection

TCP uses the Internet Protocol (IP) address and a port number on each host to create a dedicated socket pair. The socket pair is established through the use of the three-way handshake, wherein session state variables are established. The TCP application programming interface appends TCP state and control values to the data it receives from the application, forming the TCP segment (see Figure 2).

6

```
 0                   1                   2                   3
 0 1 2 3 4 5 6 7 8 9 0 1 2 3 4 5 6 7 8 9 0 1 2 3 4 5 6 7 8 9 0 1
+-+-+-+-+-+-+-+-+-+-+-+-+-+-+-+-+-+-+-+-+-+-+-+-+-+-+-+-+-+-+-+-+
|          Source Port          |       Destination Port        |
+-+-+-+-+-+-+-+-+-+-+-+-+-+-+-+-+-+-+-+-+-+-+-+-+-+-+-+-+-+-+-+-+
|                        Sequence Number                        |
+-+-+-+-+-+-+-+-+-+-+-+-+-+-+-+-+-+-+-+-+-+-+-+-+-+-+-+-+-+-+-+-+
|                     Acknowledgment Number                     |
+-+-+-+-+-+-+-+-+-+-+-+-+-+-+-+-+-+-+-+-+-+-+-+-+-+-+-+-+-+-+-+-+
|  Data |           |U|A|P|R|S|F|                                |
| Offset| Reserved  |R|C|S|S|Y|I|            Window              |
|       |           |G|K|H|T|N|N|                                |
+-+-+-+-+-+-+-+-+-+-+-+-+-+-+-+-+-+-+-+-+-+-+-+-+-+-+-+-+-+-+-+-+
|           Checksum            |         Urgent Pointer        |
+-+-+-+-+-+-+-+-+-+-+-+-+-+-+-+-+-+-+-+-+-+-+-+-+-+-+-+-+-+-+-+-+
|                    Options                    |    Padding     |
+-+-+-+-+-+-+-+-+-+-+-+-+-+-+-+-+-+-+-+-+-+-+-+-+-+-+-+-+-+-+-+-+
|                             data                              |
+-+-+-+-+-+-+-+-+-+-+-+-+-+-+-+-+-+-+-+-+-+-+-+-+-+-+-+-+-+-+-+-+
```

TCP Header Format

Figure 2.   TCP Header Format. From[5]

Port numbers tell the host that the application for which the segment is intended. The Internet Corporation for Assigned Names and Numbers (ICANN) is responsible for the oversight of protocol identifiers (port numbers), domain name spaces, and Internet address allocation. The Internet Assigned Numbers Authority (IANA), formerly responsible for this under contract to the U.S. government, maintains a listing of port numbers, known as well-known ports, used by specific applications on www.iana.org. For example, File Transfer Protocol (FTP) uses port 20. IANA is now a department within ICANN.

The host desiring to initiate a connection will use the well-known port to let the receiving host know what application to use. The process of setting up the connection is referred to as the three-way handshake. TCP is full duplex which means, once the connection is established both hosts can send and receive simultaneously.

For purposes of explanation, the host initiating the connection will be referred to as Host A and the host receiving the connection will be referred to as Host B.

Host B will initially be waiting for a connection request on any port in the *LISTEN* state [5]. Host A will determine which application to use and create the socket pair accordingly. Host A will send a *SYN* message to Host B with an initial sequence number and will be in the *SYN_SENT* state. Once Host B receives the SYN message, it enters the *SYN_RCVD* state. It will reply to Host A with the *SYN/ACK* message using the same socket pair and with its own initial sequence number. Upon sending the *SYN/ACK* message, Host B will also be in the *SYN_SENT* state. Once Host A has received Host B's *SYN/ACK* message, it enters the *ESTABLISHED* state because it has all of the state information required to communicate with Host B. Host A will reply with an *ACK* message, however, since it is in the *ESTABLISHED* state it can send data with the *ACK*. Once Host B receives the *ACK* message from Host A it also enters the *ESTABLISHED* state. Both hosts are now ready to send and receive data and may do so independently of the other host. Refer to Figure 3 TCP State Transition Diagram and Table 1 TCP State Transition Diagram for more information on the TCP state transition process.

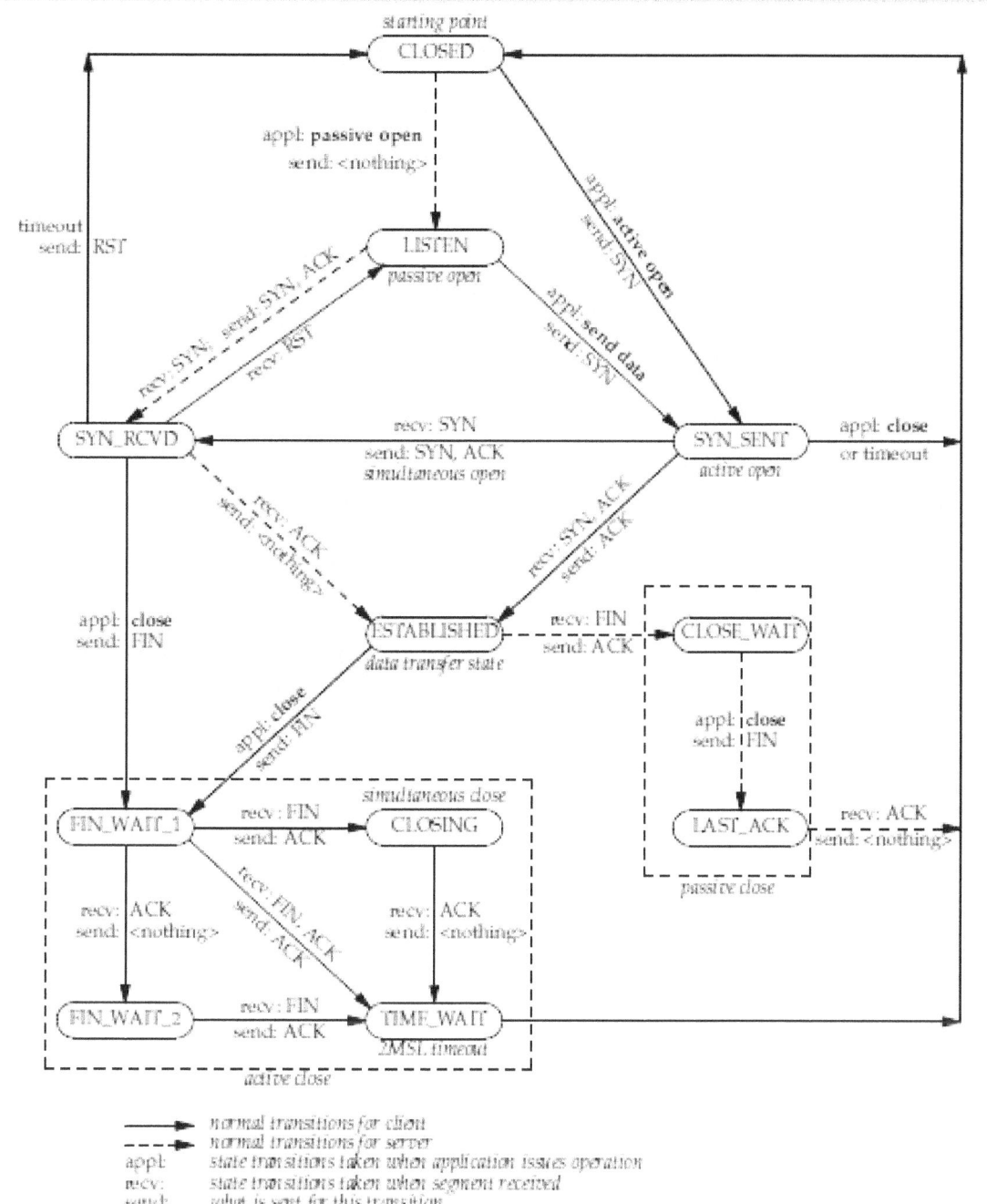

Figure 3.   TCP State Transition Diagram. From [6]

Table 1.   TCP State Transition Diagram. From [6]

LISTEN - represents waiting for a connection request from any remote TCP and port.

SYN-SENT - represents waiting for a matching connection request after having sent a connection request.

SYN-RECEIVED - represents waiting for a confirming connection request acknowledgment after having both received and sent a connection request.

ESTABLISHED - represents an open connection, data received can be delivered to the user. The normal state for the data transfer phase of the connection.

FIN-WAIT-1 - represents waiting for a connection termination request from the remote TCP, or an acknowledgment of the connection termination request previously sent.

FIN-WAIT-2 - represents waiting for a connection termination request from the remote TCP.

CLOSE-WAIT - represents waiting for a connection termination request from the local user.

CLOSING - represents waiting for a connection termination request acknowledgment from the remote TCP.

LAST-ACK - represents waiting for an acknowledgment of the connection termination request previously sent to the remote TCP (which includes an acknowledgment of its connection termination request).

TIME-WAIT - represents waiting for enough time to pass to be sure the remote TCP received the acknowledgment of its connection termination request.

CLOSED - represents no connection state at all.

When either host is ready to close the connection, it will send a *FIN* message to tell the other host that it does not have any more data to send. For this example, Host A will initiate closing the connection though either host could do so. Host A will send Host B the *FIN* message and enter the *FIN_WAIT_1* state. If Host B does not have data to send to Host A, it will reply with a *FIN/ACK* message to Host A and enters the *LAST_ACK* state. When Host A receives the *FIN/ACK* message it will send an *ACK* message and enter the *TIME_WAIT* state. The *TIME_WAIT* state is where the host sets a timer and waits a predetermined amount of time to ensure the other host received its final ACK before closing the connection. When Host B receives the final ACK message of Host A, it deletes the connection. Once the *TIME_WAIT* period has expired, Host A can also delete the connection. If Host B receives the *FIN* message from Host A and it does have data to send, it will reply with an *ACK* message to Host A and enter the *CLOSE_WAIT* state. When Host A receives the *ACK* message from Host B, it will transition from the *FIN_WAIT_1* state to the *FIN_WAIT_2* state. Once Host B has sent all the data it needs to and is ready to close the connection, it will send a *FIN* message to Host B and enter the *LAST_ACK* state. When Host A receives Host B's *FIN*, it will reply with an *ACK* message and enter the *TIME_WAIT* state. When Host B receives the final *ACK* message from Host A it will delete the connection. Host A will also delete its connection once the time period specified for the *TIME_WAIT* state has elapsed.

2. Reliable Data Transfer

RFC 793 states that TCP should provide reliable data transfer and that all data should be provided in order to the application at the end host. A TCP sender breaks down an application message (e.g., an HTTP request) into one or more data units called segments, and each such segment is wrapped in one or more IP datagrams [5].   IP does not guarantee datagram delivery, nor that they will be delivered uncorrupted and in order. Since IP performs best-effort delivery service, TCP implements a reliable data transfer using byte-oriented accountability, flow control, and congestion control. TCP uses Round Trip Timers (RTT), and cumulative acknowledgments (ACKs) for data accountability and congestion control processes.

TCP uses sequence and acknowledgment numbers to account for data. This allows for recovery when data are lost in transit, arrive corrupted, or arrive out of order. TCP assigns each byte in the segment with a 32-bit sequence number. The sequence number references an offset from the first byte of data in the segment. The receiving host sends an acknowledgment number that states it received all of the bytes up to, but not including, the acknowledgment number; thus the acknowledgment is considered cumulative and anticipatory, that is, it "anticipates" the next byte to be received.

Since each segment has a sequence number, the receiving host sends the acknowledgment for the last byte of data that it received in order. If the receiving host receives segments out of order it will use the sequence numbers to put them in order in a buffer. If there is a gap in the data, the acknowledgment number will reflect the first byte of data in the gap. For example, if Host A transmitted ten segments that each contained 100 bytes and Host B received bytes 0-499 and 600-999, Host B would set its acknowledgment number to 500. The receiving host will also use a checksum within the TCP segment header to detect if there were errors in the segment. If the receiving host determines there was an error with a segment it will drop the received segment and send a duplicate acknowledgment for the last valid segment it received.

The sending host sets a retransmit timer, if the acknowledgment is not received by the time the retransmit timer expires, the sending host will retransmit that segment. This is referred to as a Retransmit Time Out (RTO) [5]. TCP can determine the most recent round trip time (RTT) by computing the difference between the timestamp of when the segment was transmitted and the timestamp of when the ACK was received. TCP performs an average of the sampled RTTs. RFC 2988 says to use the smoothed round trip time (SRTT) formula to estimate the RTT: $NewEstimateRTT = (1 - \alpha) * PreviousEstimateRTT + \alpha * SampleRTT$ [7]. Using an exponentially weighted average for the EstimatedRTT, means TCP is continually sampling the round-trip time and using it in the average, thereby probing instantaneous network load.

There are several factors within the network that can cause the RTT to increase or decrease, such as limited bandwidth or congestion causing queuing delay or segment loss at the router. Adjusting the RTOs based on the Estimated RTTs allows the sending hosts to adapt their effective load to changing network conditions, such as an increase or decrease in congestion. The use of a small α in the weighted average reduces the emphasis on recent samples. Thus a significant change in the most recent sample does not cause a significant swing in the new RTT estimate and prevents over-reaction due to fast changing conditions. RFC 2988 also states that the retransmission timeout (RTO) value should be conservative to prevent unnecessary retransmissions; thus the timer is "padded" with a small value based on the deviation between the observed RTT and the estimated RTT. The retransmission timer will track the time of the last unacknowledged segment, i.e., the lowest sequence numbered unacknowledged byte. If the retransmission timer expires, the sender will retransmit the segment. If the sender knows that it sent segments after the last byte that was acknowledged then it may retransmit the subsequent segment, depending on the respective host implementation of the TCP protocol allows for selective segment retransmission (SACK) or simply the default to retransmit all unacknowledged segments.

There are three cases for which the sender will retransmit segments. First, the sender will retransmit when the RTO expires and the segment has not been acknowledged. Next, the sender will perform a fast retransmit when the receiver sends a duplicate acknowledgment three times. Finally, it will retransmit if the SACK option is set and a specific segment is selectively acknowledged by the receiving host. The SACK option is discussed in RFC 2018 and allows the receiver to request one specific segment rather than for the oldest outstanding segment and everything that comes after it. Figure 4 illustrates an example where a receiving host has received packets one, two, four and five but did not receive packet three (bytes 1-199, and 300-499).

Figure 4.   Missing Packet Example

Without the SACK option set, the receiving host would send an ACK for byte 200 and the sending host would retransmit packets three, four and five (starting at byte 200).   With the SACK option set, the receiving host would send an ACK for bytes 200-300 and only retransmit packet three.

3.   Flow Control

A receiving host allocates a certain amount of buffer space for a new connection.   If a sending host sends more data than a receiving host can process or buffer, the receiving host will drop some of these segments.   This results in needless retransmissions by the sending host. TCP avoids these needless retransmissions by using flow control.   Flow control is limiting the amount of data the sender transmits to prevent overwhelming the receiver's buffer. TCP uses a "receive window" field in the TCP header of each ACK to specify the amount of additional data the receiver is able to receive.   This lets the sending host know how much more data it can send.   If the receive window goes to zero, the sender will periodically probe the receiver to see if the receive window has opened back up by sending a 1 byte segment [8].   This way the communication

15

between the receiving function and the corresponding send function is kept open, allowing the receiver to process data in its buffer while the source is throttled until notified that it can resume sending more segments.

4. Congestion Control

Congestion control is limiting the amount of data transmitted into a network. Congestion on a network can cause segments to be delayed or dropped by routers along the way to the end host. TCP needs to know how to deal with this to prevent needless retransmissions thereby adding to the congestion problem. TCP tries to minimize this effect by determining network conditions and transmitting at a rate that is as high as possible while minimizing segment loss. It uses end-to-end congestion control which means it uses the end hosts to perform congestion control and not the routers or other network devices. TCP assumes all segment loss is caused by congestion and responds by reducing the send rate. TCP's congestion control algorithm is specified in RFC 5681 (see Figure 5).

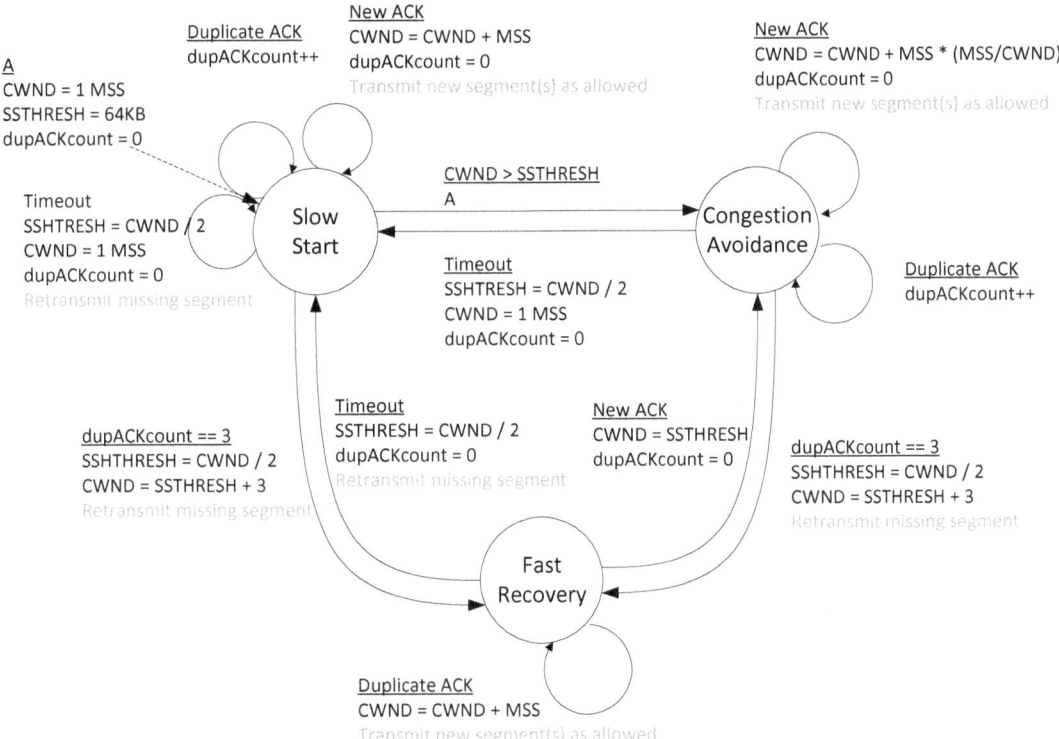

Figure 5.   TCP Congestion Control Algorithm. After [8]

a.   *Slow Start*

TCP uses the slow start algorithm to probe the network to determine congestion conditions and to limit sending too much data into a congested network. It manages a derived congestion window (*CWND*) to control the amount of data the sender injects onto the network as a means of establishing the effective network bandwidth. RFC 5681 states that *CWND* is the sending host's limit on the amount of data it can send without receiving an acknowledgment. To be more specific, the send rate is not to exceed the *CWND* or the receive window (*RWND*), whichever is less. If the receiving host's receive buffer is full and has a receive window of 1 but *CWND* is higher, the sending rate is reduced to the receive window. Understanding that TCP takes both into account, the following discussion will assume a large *RWND* and discuss the sending rate using only *CWND* in order to focus on how it implements slow start and congestion control.

CWND starts at one Maximum Segment Size (MSS), as determined by the session initiation handshake, and doubles with each successful RTT until it reaches the slow start threshold (*SSTHRESH*); that is, for each segment successfully acknowledged, the source opens its congestion window by another segment, effectively doubling the size of the congestion window when all sent segments are acknowledged. *SSTHRESH* is used to determine when TCP transitions from slow start to congestion avoidance as it determines the effective transmission rate [9]. Once the sending rate reaches *SSTHRESH*, TCP transitions from slow start to congestion avoidance. If TCP detects a lost segment through an RTO, then it reduces *CWND* to one and reduces *SSTHRESH* to the *CWND* divided in half and starts over in slow start.

b. Fast Retransmit/Fast Recovery

If the receiving host receives a segment out of order, it will immediately send a duplicate acknowledgment for the last segment it received in order [9]. If the sending host receives three duplicate acknowledgments, it will perform a fast retransmit and retransmit the missing segment (see Figure 6).

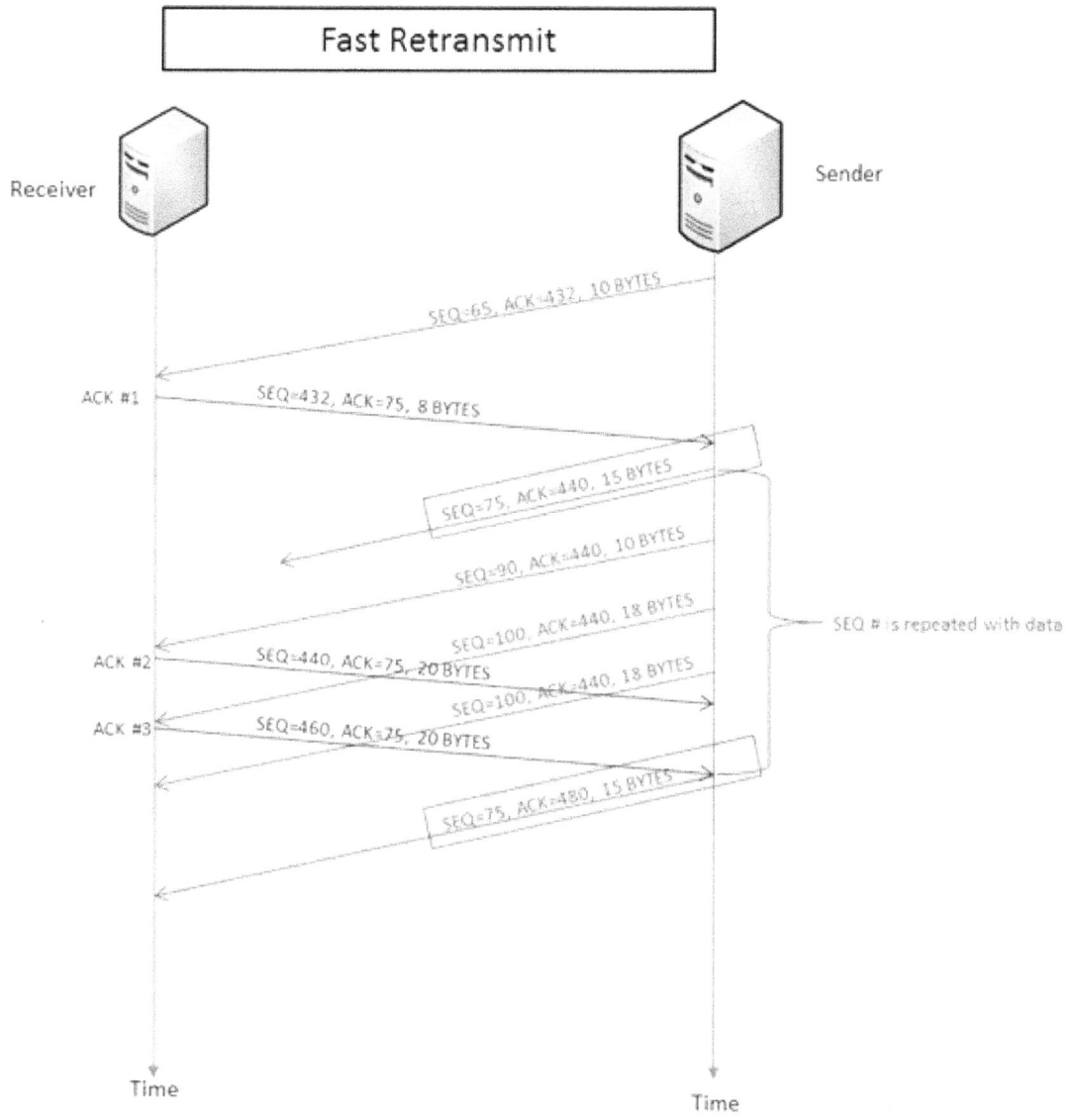

Figure 6.   Fast Retransmit

After performing a fast retransmit, the sending host will enter fast recovery. Fast recovery covers the period from the fast retransmit until the sending host no longer receives duplicate acknowledgments from missing data. In fast recovery, if the TCP implementation is TCP Reno, *SSTHRESH* is set to ½ *CWND* and the sending host continues into congestion avoidance. If TCP Tahoe is being used, then the *CWND* is set back to 1 MSS [8].

c. Congestion Avoidance

During slow start the sending rate is rapidly increased until congestion is detected (TCP Reno) or *SSTHRESH* is reached. Once slow start detects congestion or reaches *SSTHRESH,* it enters congestion avoidance. In the former case, *CWND* is set to half of its previous value if TCP Tahoe is used. In either case, *CWND*, increases by one MSS with each successful RTT while in Congestion Avoidance.

If there is a RTO while in congestion avoidance, *CWND* is set to 1 again and *SSTHRESH* is set to half of *CWND*'s last value and starts over in slow start. If there is a triple ACK, which triggers a fast retransmit, *CWND* is reduced to half its value, *SSTHRESH* is reduced to half of the value of *CWND* (see Figure 7) [8].

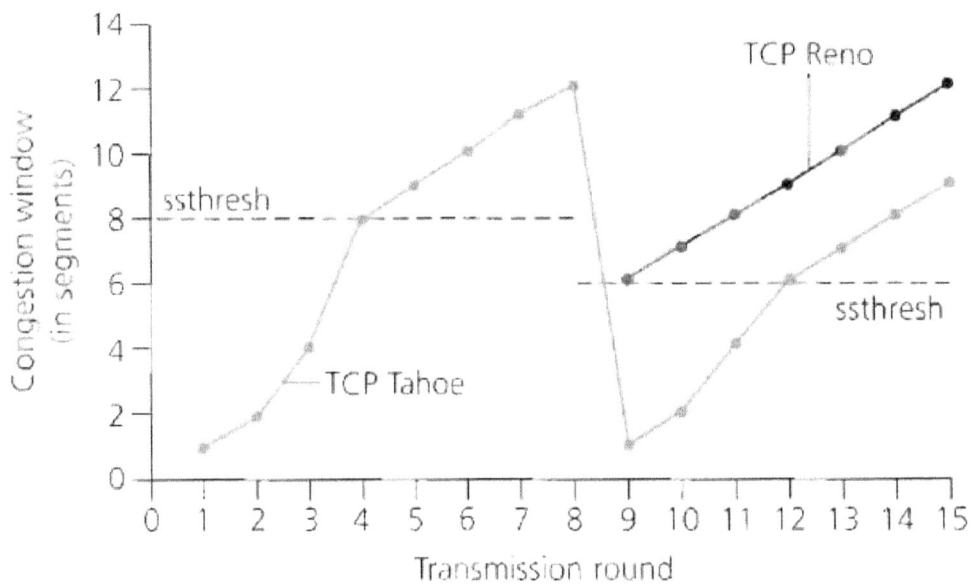

Figure 7.   TCP Congestion Control [8]

20

5.   TCP Message Format

The TCP header contains a source and a destination port number, sequence and acknowledgment numbers, flags, and options. The sequence and acknowledgment numbers, along with the flags, form the state information. The MSS is the largest amount of data that can be put in a segment and does not include the header for TCP or IP [10]. The MSS is determined by how much data can be encapsulated in the link layer frame. The IEEE 802.3 frame, colloquially referred to as an Ethernet frame, allows up to 1460 bytes of application data, assuming 20 bytes each for the TCP and IP headers. This is usually the largest amount of data for an MSS but RFC 1323 [11] provides an option for sending more data.

As part of the three-way handshake initiating a TCP session, the host initiating the session chooses an arbitrary sequence number. RFC 793 states each host should use an initial sequence number generator using a 32 bit clock, however it could be implemented differently [5]. If initial sequence numbers started from 1 or some other predetermined point there could be a problem if a connection using the same application between two hosts is initiated, closed, and then restarted within a short period of time. A host would be unable to determine if the segments received were from the first connection that was closed or if they came from the recently initiated connection [5]. It also prevents a malicious third party from guessing the sequence number and sending false segments, performing a TCP Sequence Prediction Attack [12] (see Figure 8).

TCP Sequence Number Starting at 0

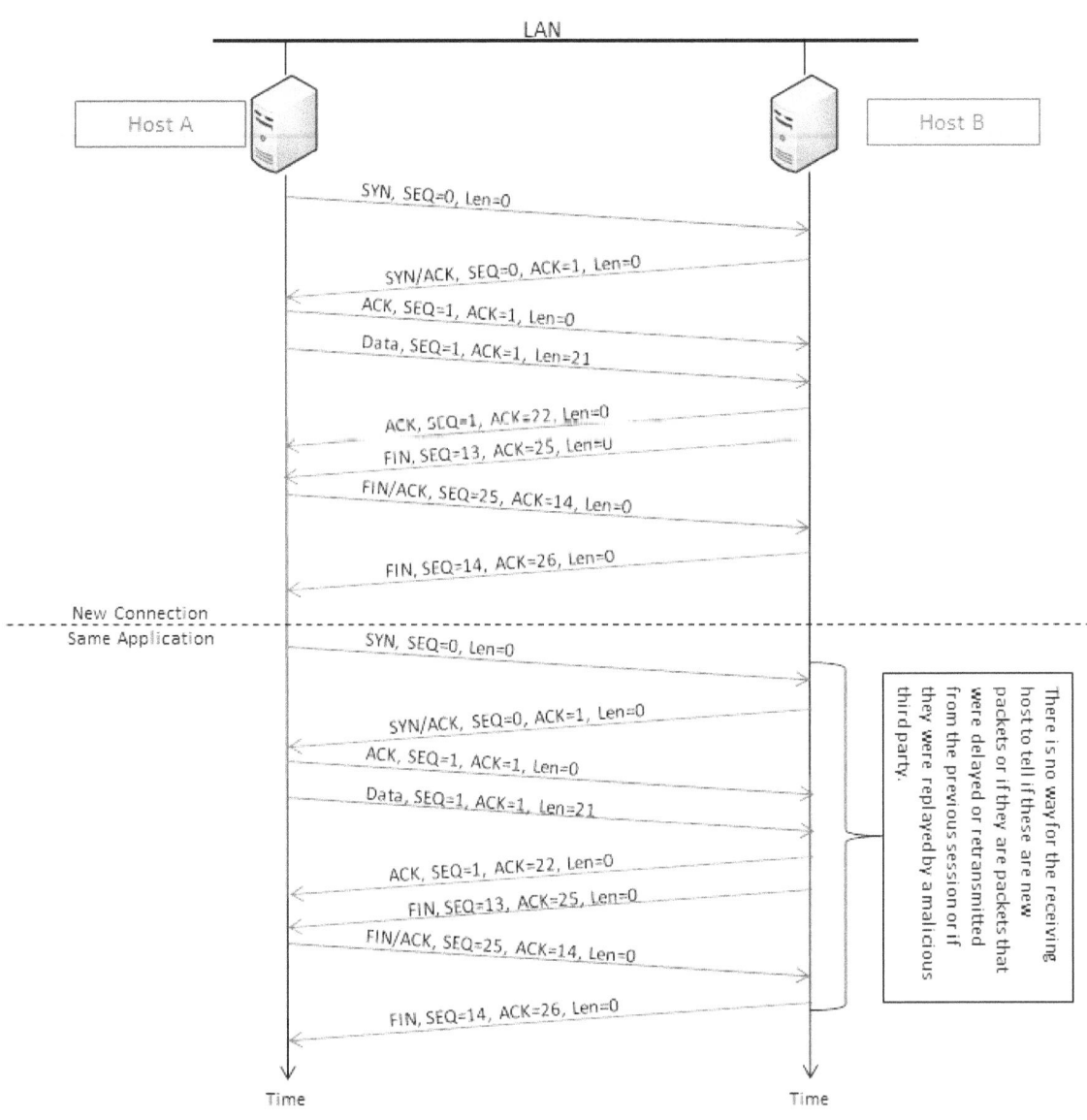

Figure 8.   TCP Sequence Numbers

From this point, the sequence number references the first byte of a segment data field to the first byte of data in the session [8]. The acknowledgment number (ACK) is used by the receiving host to indicate the next byte it expects to receive; the acknowledgment number is referenced to the sequence number of the sending host. As noted previously, TCP uses anticipatory, cumulative acknowledgments, which means the acknowledgment number confirms to the recipient that all bytes sent up to but not including the byte corresponding to the acknowledgment number have been successfully received by the acknowledging host. If the receiving host receives a segment that is out of order, it will acknowledge the last byte it received in order up to the missing byte by sending the sequenced number of the byte it expected (see Figure 9).

TCP Sequence Number Starting at 0

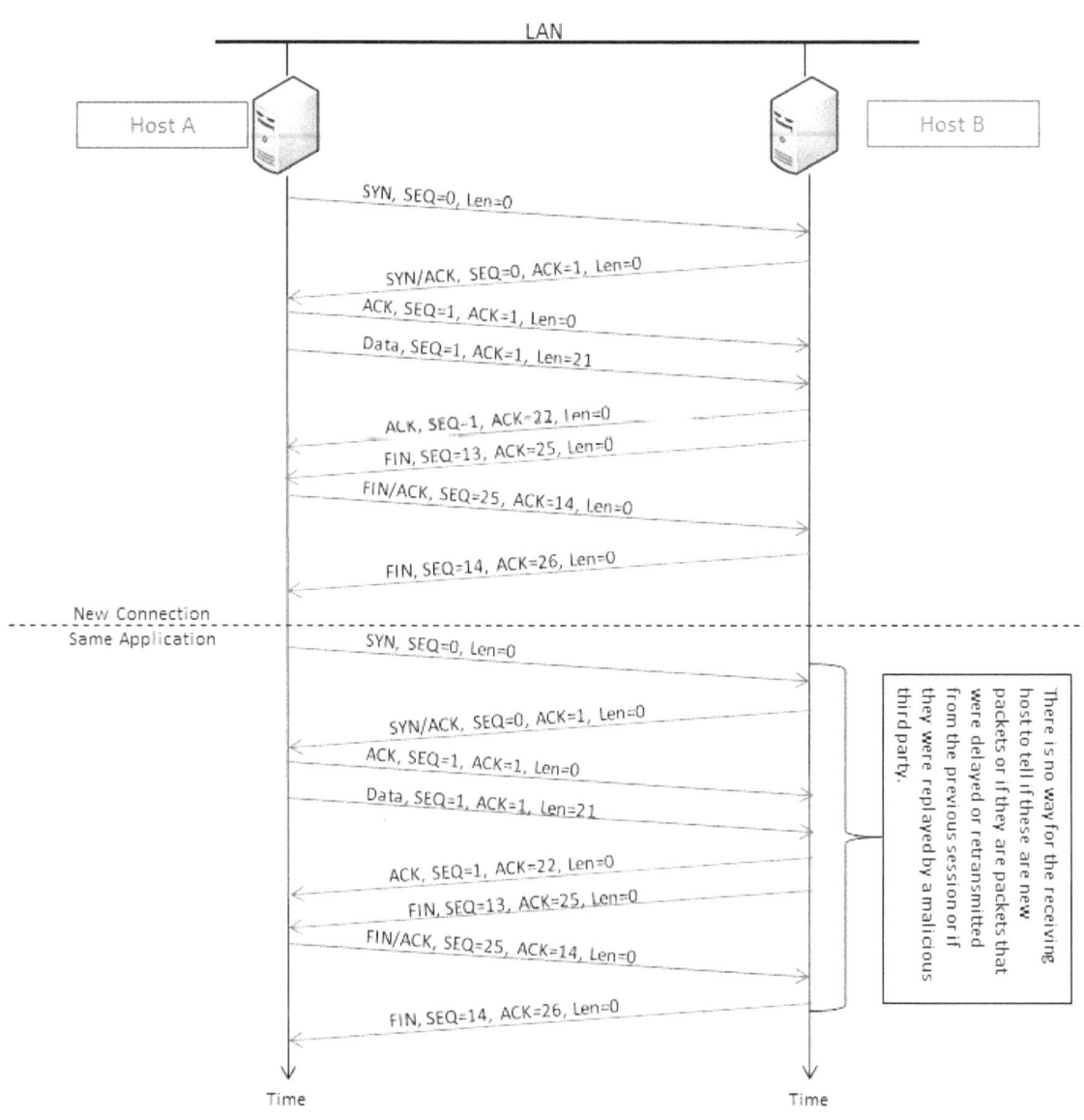

Figure 9.   TCP Sequence Numbers

The flag field contains six bits. If the corresponding bit contains a 1, then the field is set. If it contains a 0, then the field is not set. The ACK field is set when the sending host is acknowledging data received, thus indicating that the value in the acknowledgment field is valid; otherwise, the value is ignored. The Reset (RST) flag is set when the sending host is requesting to reset the TCP connection. The SYN flag is set when the sending host is trying to open a TCP connection; the respective header includes the set-up state information of the sender. The FIN flag is set when tearing down the TCP connection and means that the sending host will not send more data and wants close the connection. The window field is used to tell the receiving host the amount of data that the sending host is able to receive. If the window field is equal to zero then the sending host cannot receive any more data and the other host will periodically probe the sending host with a segment with one byte of data to see if it is ready to receive more data.

RFC 1323 specifies the options that may be in the options field. Important options are the window scaling option and the Selective Acknowledgments (SACK) option. The window-scaling allows the sender to send data larger than the MSS would otherwise allow, if the recipient Host Agrees. The SACK option allows the receiving host to send SACK packets to the sender informing the sender of the all the data that has been received, not just up to the first byte missing. This allows the sender to just retransmit the data that was missing [14].

B. WAN OPTIMIZATION DESCRIPTION

WAN Optimization deals with making data transfers across a Wide Area Network (WAN) more efficient. The mechanisms that make TCP reliable for normal file transfer make it inefficient for file transfers where there is a long RTT. When TCP has a very long RTT it is possible to send enough data that the sending host fills up its send buffer and has to wait for an ACK from the receiving host before it can send more data. As a result, even though there is more bandwidth available, the sending host has to stop and wait until it receives an ACK before it can continue sending. In other words, the sending host's send window becomes the bottleneck rather than the bandwidth of the link. This is often referred to as the Long Fat Pipe (LFP) or Long Fat Network (LFN) problem and RFCs 1072 [13] and 1323 propose mechanisms to overcome the issues presented by an LFN.

The bandwidth-delay product (BDP) is a good measure to determine if a network is an LFN. The BDP is the bandwidth * RTT and it gives a measure of the capacity of the link, or how many unacknowledged bits can be in transit on the link at one time. RFC 1072 says that any network that can have 10^5 bits in transit at one time is an LFN. If the number of bits currently in transit on the link are greater than the send window of the transmitting host, it will be forced to stop and wait for ACKs before it can continue sending which effectively reduces its send rate.

Optimizing data transfers is needed when the network has a very long RTT. A common example of an LFN would be a network with limited bandwidth connected via satellite where the extreme distance causes a very long RTT due to propagation. There are several topologies for WAN optimization but the most prevalent is where the client sends its TCP segments to an optimizer, which then performs the optimization and forwards the segments on the link. On the receiving side, the receiving host's optimizer receives the segments sent by the sending host's Optimizer and then forwards them to the receiving host (see Figure 10). To be clear there is usually one WAN optimizer for each LAN segment on a WAN. The WAN Optimizer is usually placed just before a long link like a satellite connection.

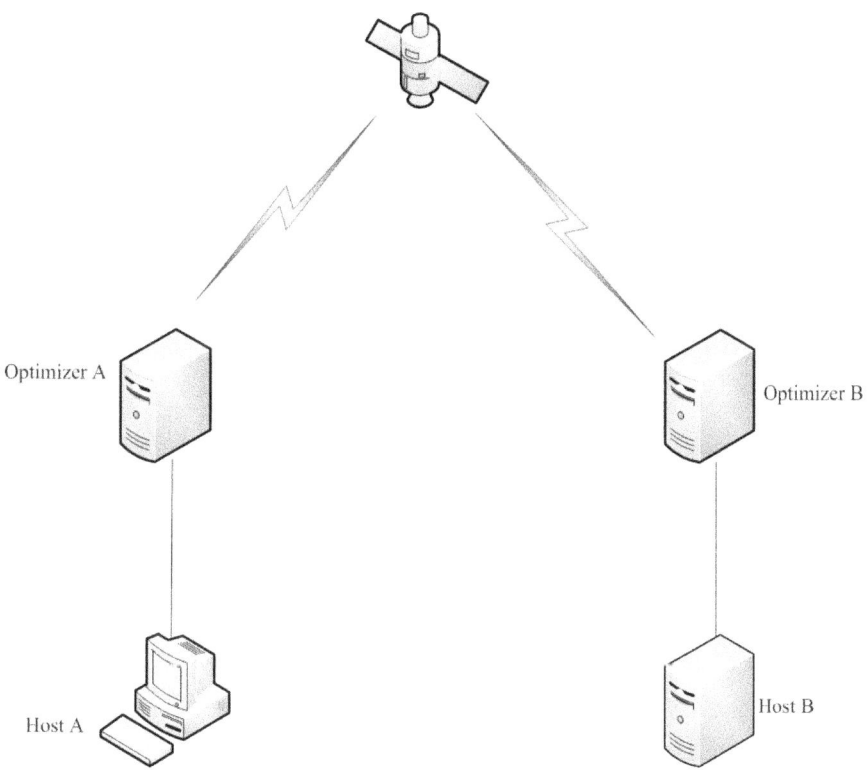

Figure 10.   Common WAN Optimization Topology

1. Methods of WAN Optimization

a. *TCP Acceleration*

When transmitting over a link with a very long RTT, TCP will have a low sending rate because the long RTT prevents TCP from opening up the send window quickly in slow-start, as described earlier. TCP acceleration attempts to keep the transmission rate high by using a proxy on each side of the connection (see Figure 11 TCP Acceleration).

Optimizer Timing Diagram

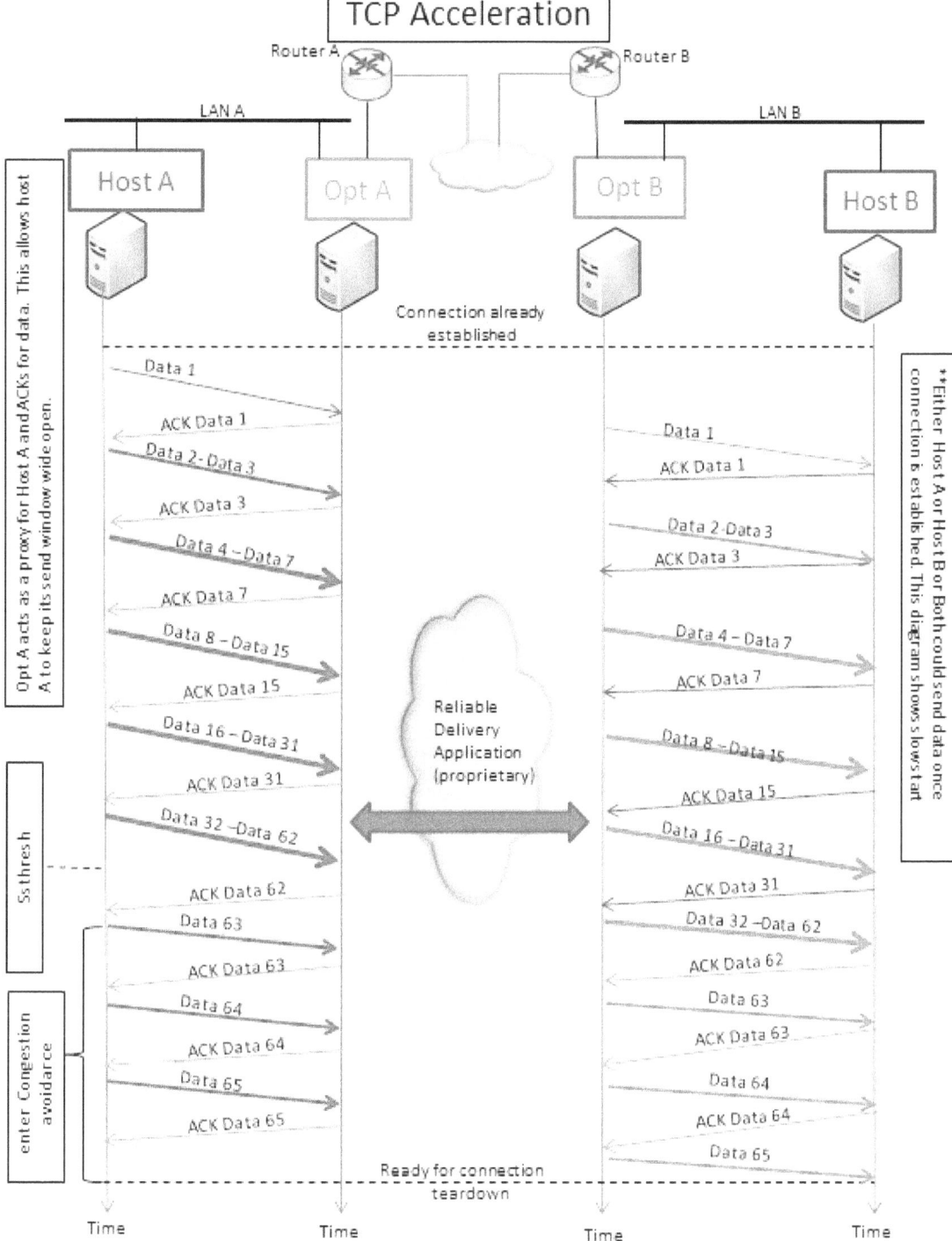

Figure 11.   TCP Acceleration

The proxy (Optimizer A for Host A and Optimizer B for Host B) acknowledges all of the hosts segments and forwards them to the other proxy via a proprietary reliable delivery application over the troubled connection, possibly using forward error correction to ensure reliable delivery of segments and minimize retransmissions across the constraining link.

The end hosts are unaware that their segments are not sent directly to the other host. As a result Optimizer A and Optimizer B must keep track of all state variables for both hosts. Figure 11 shows how the optimizers store the state data needed for each host during the TCP three-way handshake. Optimizer A cannot ACK the SYN for Host A because it does not know if Host B is there or will accept the connection or what sequence number Host B will use. Optimizer A has to relay the SYN to Optimizer B and Optimizer B will relay the SYN to Host B. Host B will send its SYN/ACK to Optimizer B. Optimizer B relays the SYN/ACK to Optimizer A which then relays it to Host A. Host A now has everything it needs to communicate through the optimizers to Host B. When Host A sends the ACK to Optimizer A, the optimizer then has all the state data it needs to act as a proxy for Host A. Optimizer A relays the ACK to Optimizer B, which then has all the state data to act as a proxy for Host B. Once the three-way handshake is complete the hosts are ready to begin transferring data.

Each time Host A sends data, Optimizer A immediately acknowledges it and then forwards it over the connecting link to Optimizer B. This allows Host A to take full advantage of the slow start algorithm to maximize the send window. Figure 12 shows a TCP connection that has a long propagation delay and high loss that does not use TCP acceleration.

Troubled TCP Connection
Without Optimization

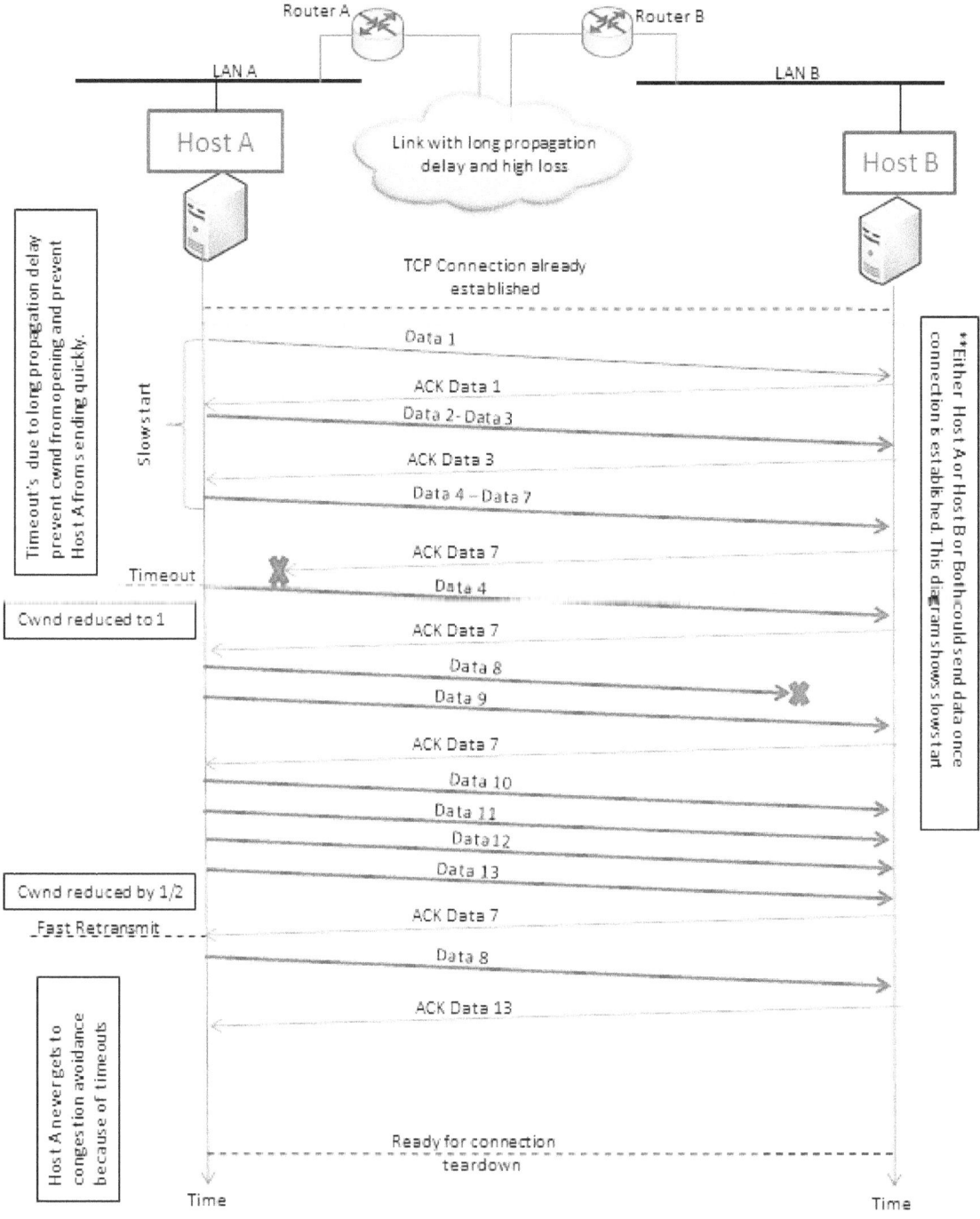

Figure 12.   TCP Without Optimization

The high segment loss prevents Host A from ever reaching *SSTHRESH* and keeps the transmission rate artificially low. It is important to note that for simplicity both Figures 11 and 12 show data flowing from Host A to Host B but TCP is full duplex and data can flow in both directions.

There are several situations that can cause an RTO. First, if the ACK segment arrives and is corrupt, it will be dropped and the retransmit timer will eventually time out. Second, if there is an error and the segment never reaches the receiving host or if the transmitted segment is received but the *ACK* message is lost, the retransmit timer will timeout. Lastly, if the RTT takes too long, the receiving host's ACK may not arrive before the RTO timer expires. All of these factors will prevent the sender from achieving an effective window size.

The teardown of the connection will occur in a similar manner to the setup. The optimizers relay the FIN and FIN/ACK messages because otherwise one of the hosts might delete the connection when the other host still has data to send. For example, when Host A sends the FIN and it is relayed to Host B through Optimizer A and B. If Host B still has data to send it would send an ACK rather than a FIN/ACK and then continue to send the last of the data it had to send. If Optimizer A had responded to Host A's FIN with a FIN/ACK, Host A would delete its connection and would not be able to receive any of the data that Host B still had to send.

b.   *Data De-duplication*

On a link that has extremely long RTT if transmission time can be reduced, the RTT will also be reduced. Data de-duplication attempts to minimize this issue by having the optimizers cache data when it is seen for the first time. If the data is sent again, the optimizer will only send the changes in the data and a token referencing the data across the congested link with the long RTT. If there are less bits being transmitted, then it will reduce the time to transmit and receive them. The end result is less data was transmitted across the bottleneck link but the receiving host receives same amount of data in less time.

In data de-duplication when Host A sends Optimizer A data, the optimizer will cache that data, assign it a token, and then forward the segments to Optimizer B. If Optimizer A receives data from Host A that it recognizes as duplicate data, it will only forward the token referencing the original data to Optimizer B; Optimizer B will then forward the full data to Host B. This reduces the time to transmit the same amount of data and allows for increased bandwidth utilization. Figure 13 shows an example of data de-duplication.

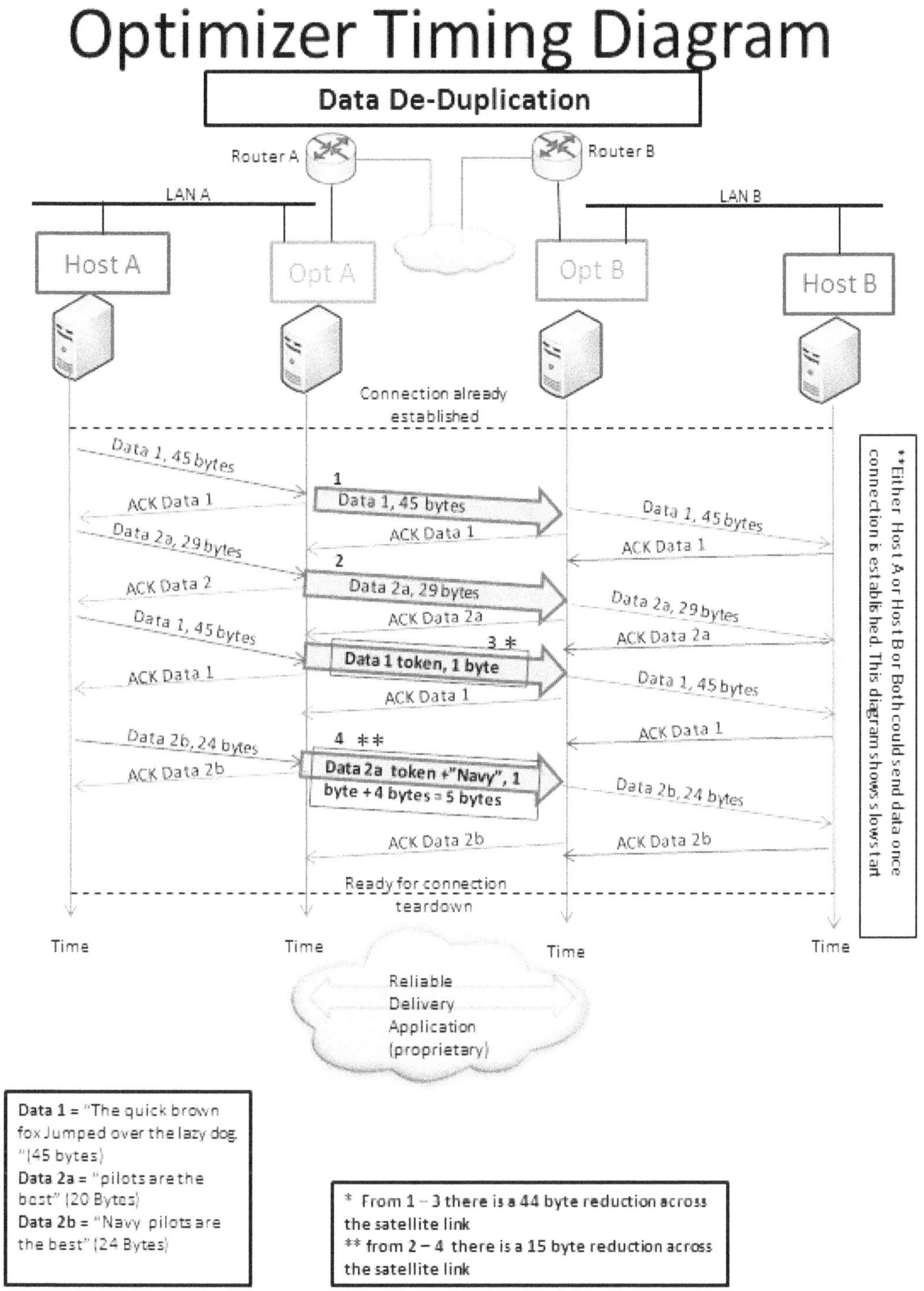

Optimizer Timing Diagram

Data De-Duplication

Figure 13.   Data De-duplication

Host A sends a 45 byte segment with the data "the quick brown fox jumped over the lazy dog" to Optimizer A.

This is the first time that Optimizer A has seen this data so it caches the data and forwards the data with the token to Optimizer B. Optimizer B also caches the data and forwards it Host B. This adds a slight amount of overhead because the original data and tokens must traverse the troubled link. However, data de-duplication greatly reduces the data transmitted when the same data repeatedly crosses the troubled link. The next time Host B tries to send the same data to Optimizer A, Optimizer A will check the cache, see the same data and forward a 1 byte token referencing the data to Optimizer B. Optimizer B receives the token, checks the cache and forwards the full 45 byte data segment to Host B.

In the next data set, Host A sends Optimizer A a 20 byte segment with the data "pilots are the best." Optimizer A caches the data and forwards the full 20 byte data segment and token to Optimizer B. Optimizer B also caches the data and the reference token and then forwards the data to Host B. Next, Host A sends a 24 byte segment with the data "Navy pilots are the best" to Optimizer A. Optimizer A checks the cache and sees that "Navy" is the only change from the previous segment that it received. Optimizer A forwards "Navy" and a 1 byte token referencing the other data to Optimizer B. Optimizer B receives "Navy" and the token, retrieves the data from cache that the token references and forwards "Navy pilots are the best" to Host B. Data de-duplication will see greater data reductions when the files are larger. For example, if the Host A initially sends a 200KB word document to Host B for storage the first time Optimizer A sees the data it will forward the data to Optimizer B and assign tokens to the data.

Optimizer B will receive all the data and the corresponding tokens and forward the data to Host B for storage. Now, let's say the user at Host A discovered an error in the document, he misspelled his name. Normally this would require, the entire 200 KB document to be retransmitted. Instead, Host A sends the document to Optimizer A, Optimizer A sees the only new data is the name and sends tokens representing the rest of the data to Optimizer B. If TCP uses a maximum segment size of 1460 bytes, it would take 137 segments to send the 200KB document. Sending 137 segments reliably over a link with long propagation delay, limited bandwidth, and high loss would take considerable time. With optimization, if there were few or no changes to the original document, it could be sent in 1 segment referencing the tokens. This causes a tremendous reduction in the data that needs to be sent and the time it takes to transmit it. Once again the end hosts are unaware that there was any change in data that was sent or received. This reduction in transmission time and the reduction in the amount of data traversing the bottleneck link makes the transfer of data across the WAN more efficient.

c.   Application Acceleration

Application acceleration is giving precedence to certain types of traffic. For example, precedence could be given to the TCP three-way handshake; thus if a SYN, SYN/ACK, or ACK segment were received at the optimizer they would be forwarded out ahead of segments already in the transmit queue. Precedence also allows the user to give priority to certain types of traffic. For example, in a military environment it might be more important to download the image of a map than it is to check email. The segments for the map download would get forwarded out of the optimizer ahead of the email segments. These application preferences would be set on the optimizers and could be tailored to the types of traffic upon which the users put the highest priority. They could also be applied to groups of users. These changes could also be applied to user groups, where it might be important for leaders to have access to email but not for subordinates.

d.   Compression

Compression involves using algorithms to encode the data so that it uses fewer bits than the original representation. Most compression algorithms are based on a variant of a Lempel Ziv compression algorithm. Compression algorithms can be described as lossless or lossy. Lossless compression algorithms ensure that none of the original data is lost. One would not want to transfer a text file and lose data in the process, it would defeat the point. Gzip and WinZip are common lossless compression algorithms [15]. Lossy compression algorithms attempt to increase the compression ratio by stripping away data that is not essential to the meaning of the content. For example, the compression algorithm used for .jpg photos removes some of the colors that are not detectable by the human eye. This causes a loss in quality of the photo but is usually difficult to notice and results in less data being used to represent the picture.

Lossless compression is most often used for optimization because it ensures all of the original information can be exactly reproduced when the data is decompressed [15]. Lossless compression algorithms look for patterns in files; when the pattern is repeated it can be referenced by a smaller construct instead of repeating the data. Such constructs may be references to a dictionary (or an indexed list) that hold the original pattern. To be effective, the references are shorter than the original pattern. In text files, the pattern might be repeated words or letters. In a picture file, the pattern might be patterns of colors or clusters of pixels. By replacing the pattern with a shorter reference, the amount of data that is transmitted may be significantly reduced.

e. Suppression

Suppression of segments involves restricting the transmission certain types of segments. There several different methods to perform segment suppression. One method is to use Selective Acknowledgment (SACK) segments rather than cumulative ACKs for each segment received. TCP SACK options are discussed in RFC 2018. By using the SACK option the sending host only retransmits the segment that was missing and suppresses retransmitting the subsequent segments. The RFC also states that if there are multiple missing segments that up to three segments may be selectively acknowledged at the same time. This would suppress duplicate acknowledgments. For example, if the sending host transmitted eight segments of 1000 bytes each and the 2nd, 4th, and 6th segments are lost or corrupted in transit. In normal TCP, when the 3rd segment is received the receiving host would ACK for byte 1001. It would ACK for the same byte when packet 5 arrived and again when packet 7 arrived. The third duplicate ACK would trigger a fast retransmit and packets 2-8 would be retransmitted. If the SACK option was enabled by both hosts when the third packet arrived, the receiving host would ACK for bytes 1001-2000. When the fifth packet arrives, the receiving host would ACK for bytes 3001-4000 and 1001-2000. When the seventh packet arrives, the receiving host would ACK for bytes 5001-6000, 3001-4000, and 1001-2000. The end result is suppressing the retransmission of segments that were already received (3 and 5).

f. Increasing TCP Active Window Size

Adjusting TCP active window size is discussed in RFC 1323. This option must be set in the SYN and SYN/ACK segments between the sender and receiver, that is, during the initial TCP three-way handshake during session setup. With this option set, the TCP send and receive windows can increased beyond the standard window size of 65535 bytes. This allows more bytes to remain in transit at one time without the sending host having to stop and wait for acknowledgments and thereby reducing some of the effects of an LFN. This capability mitigates the potential upper-bound constraint placed by the receive window on the congestion window size.

2. PERFORMANCE EVALUATION OF WAN OPTIMIZERS

Any method used to make the transfer of data across a WAN more efficient can be referred to as WAN optimization. WAN optimization is offered by several different vendors and each one has proprietary methods of performing optimization; however, most use some or all of the following methods: TCP acceleration, data de-duplication, application acceleration, compression, suppression of unnecessary segments, and more efficient use of high speed WAN links by scaling up the TCP receive window size. Many of these WAN optimization vendors claim to increase performance but the question remains as to how the claims should be tested?

One way would be to execute a quantitative measurement by performing a data transfer over a satellite link. This would be ideal because it is real TCP traffic and has real data in the payload. It also represents the timing between segments that is very difficult to re-create in a lab setting. Further, it maintains statefulness since TCP establishes a connection for the data transfer. However, this is costly and potentially disruptive to test the performance of WAN optimizers in deployed networks.

Alternatively, it may be cost-effective and less costly to evaluate the performance of WAN optimization products in a lab setting using traffic traces collected from deployed networks. This approach requires that other factors be considered. For example, a local network can simulate long delay RTT and set specific transmission error rates that simulate some of the metrics observed in traffic patterns on the Internet; however, some realism will be missing. Simply replaying the packet traces in a lab setting cannot re-create the all of the interactions with the various routers and other network devices that impact the segments as they are forwarded to their ultimate destination. Routers can drop, delay, or fragment segments based on how they are configured and depending on the traffic conditions they are experiencing. Testing in a lab environment gives a good general idea of what will happen but it must be taken into account that real traffic will behave differently. As a result, testing WAN performance in a lab environment must be done very carefully, testing a specific performance metric and not attempting to be too general.

A WAN emulator is a networking device that allows the user to artificially impose network conditions, like Bit Error Rate (BER), propagation delay, and bandwidth, experienced by a particular transmission system. For example, a satellite system on the back of a military truck might have a different BER and bandwidth than a commercial satellite system on a building. The propagation delay would vary with distance from the satellite and the end host, most specifically impacted by the angle of acquisition by the ground station or user terminal. A user can take measurements on an actual network configuration being used and then apply those to the WAN emulator. WAN performance can be tested by creating a simulated WAN using a WAN emulator and end hosts, while tailoring network performance metrics like the BER, bandwidth, and propagation delay to the metrics of interest.

Using a WAN emulator and WAN optimization devices, a simulated WAN can be created with network metrics that closely match WAN metrics observed or measured in the real world. Such a setup could be modified in many ways to approximately match many real world WAN setups. WAN performance can be tested by creating web or file transfer servers and then measuring the time it takes to complete a transfer across the network. This test would maintain statefullness and use real data but would lack some accuracy because these tests are conducted on a simulated WAN and not a real WAN. The results are usually seen as a best case scenario because the networking equipment is only handling the traffic that is generated during the test and not all the traffic that normally flows over a real WAN.

A transfer may be described as either cold or warm, depending on whether the optimizer being considered has encountered any of the data previously. A cold transfer is where the optimizers have not seen any of the data. A cold transfer can test TCP acceleration, application acceleration and compression but not data de-duplication. A warm transfer is where the optimizers have seen the data before and are able to perform data de-duplication. A warm transfer should see a significant reduction in both the amount of data transferred and the time it takes to complete the transfer.

WAN performance can also be tested by sending TCP segments with an artificial payload on a WAN emulated network. This allows for stateful replay which allows congestion and flow control to be measured. However, an artificial payload does not allow compression or data de-duplication to be tested. If the payload is filled with a repeated character or all of the segments contain the same payload then compression and data de-duplication would test artificially high. If random data is put into the payload then they would not be able to perform compression or data de-duplication and the test results would be artificially low.

Another way to test WAN performance is to use a realistic payload without stateful TCP session establishment using a WAN emulator in a lab setting. This is done using tools such as *tcpreplay*, which takes TCP segments from a network trace .pcap file and replays them in order. This would maintain the timing of the Internet and use real data if the segments were replayed with the same timing sequence. However, *tcpreplay* replays the segments at the data link layer and not the transport layer. It also plays all of the segments from the same source and does not look at the sequence or acknowledgment numbers and does not perform the setup or teardown of the TCP connection; therefore it does not maintain state appropriately. As a result, the optimizer does not see the traffic as TCP traffic and is unable to perform any optimization.

The purpose of this thesis is to create a tool to test WAN optimization by performing a stateful TCP replay of a live network trace to test the efficiency of WAN optimizers. The tool would utilize a live network trace file of a TCP flow between two hosts performing a file transfer and split it into two trace files based on the sending host. Each host would create a queue of packets based on its trace file and replay them based on the sequence and acknowledgment numbers. This would re-create the setup and teardown of the TCP session and allow the optimizers to track all TCP state variables for the entire file transfer.

III. METHODOLOGY

A. REQUIREMENTS/GOALS

Marine Corps Tactical Systems Support Agency (MCTSSA) located at Marine Corps Base (MCB) Camp Pendleton, CA is performing evaluations of WAN Optimization equipment and has requested a solution that will have the following features [16]:

- MCTSSA Problem Statement: "MCTSSA is interested in the research of a tool used to create repeatable, realistic, and stateful network tests using actual data captured from live networks."

- Act as client and server in a TCP exchange.

- Read in a standard .PCAP file that is not severely limited by size (>50GB).

- Parse packets and separate into flows while removing TCP errors and retransmissions.

- Modify packet headers to represent the test scheme.

- Replay TCP data maintaining inter packet timing and a stateful TCP stack

1. Scope

The goal of this thesis is to lay the groundwork to provide a tool with the features MCTSSA has requested. The tool will focus on replaying a single two-way captured TCP data flow between two hosts in a laboratory setting. The replayed TCP flow should maintain the relative ordering of packets and their TCP payload data while removing retransmissions. The prototype should also ensure that tests are repeatable with minimum overhead and that different TCP flows can be selected and replayed in order.

Due to time and equipment restraints, the tool created in this thesis will not be able to accomplish all of the requirements stated by MCTSSA. This prototype will only test one TCP flow at a time and ensure packets are replayed in order with retransmissions removed. However, the tool was configured so that it could be expanded to do multiple flows at a later time. Trace files above 50MB strain the resources of the computers used to perform the testing. If initial testing is successful, then larger trace files can be tested later using computers with more resources. Retransmissions should be removed because the final goal is to test the ability of WAN optimization products to make a TCP flow more efficient. If retransmissions were left in the replayed TCP .pcap it would artificially raise the amount of work the optimizer tools were trying to perform.

B. DESIGN

The prototype consists of two Python programs: tcpFlowPrepper.py and orderedReplay.py. Figure 14 shows the architecture of the tool. The first program, *tcpFlowPrepper.py*, will read in a live network trace file and put all of the IP packets into a MySQL database. The program will then query the database to check all the TCP packets to see if they are retransmissions and, if they are, mark them so they will not be used later. The program must be able to query the database for a specific TCP flow and split that flow into two .pcap files based on the source IP address. The first program will also generate a timing file for the second program to use to determine if the host running the program should send a packet or listen for a packet.

The second Python program, *orderedReplay.py*, will re-create a TCP flow between two hosts. This program will be run on two hosts and is responsible for reading in the appropriate .pcap file, resending the TCP packets from it in the correct order, and listening for packets from the other host (see Figure 14). The order of transmission is determined by the text file that was created by *tcpFlowPrepper.py*. The text file contains the name of the host transmitting and a number representing the number of times the host should transmit before pausing to receive packets (see Figure 15).

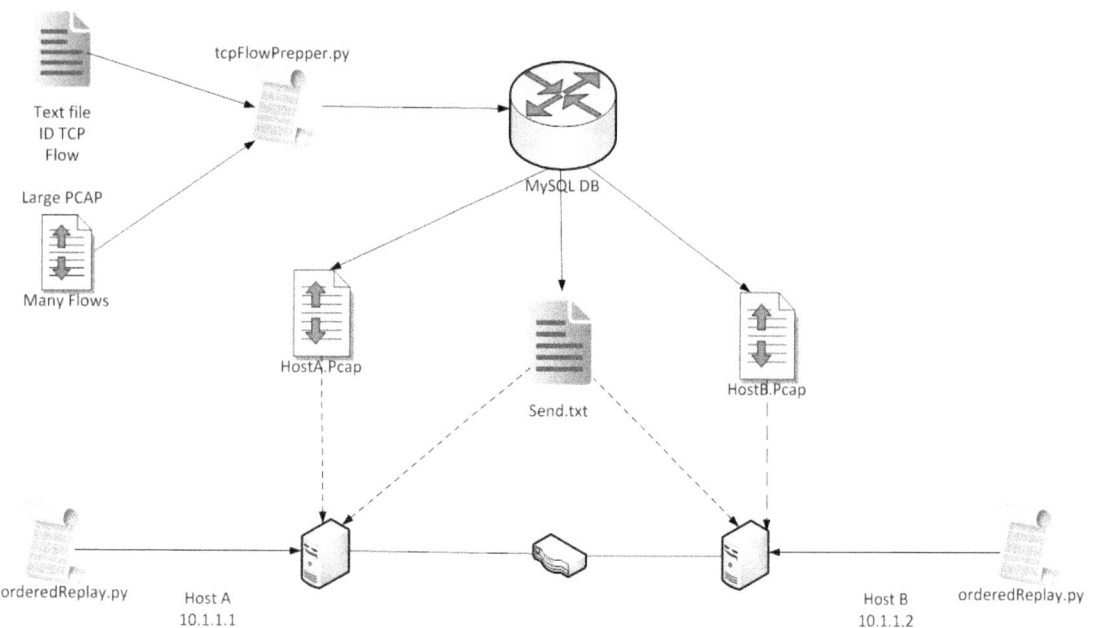

Figure 14. Program Architecture

```
hosta 1
hostb 1
hosta 2
hostb 2
hosta 1
hostb 1
hosta 1
hostb 1
hosta 3
hostb 2
hosta 3
hostb 2
```

Figure 15.   Sample order.txt

Figure 16 shows the test configuration with capture
points.   To test the tool, a network trace would be taken
when running the test and compared to the original network
trace in Wireshark with a filter applied for the targeted
TCP flow.   Noted differences for the new trace should be
that retransmitted packets are removed, and the MAC and IP
addresses are changed to match the laboratory testing
scheme.   Packet ordering, TCP source and destination port
numbers, sequence and acknowledgment numbers, flags,
receive window size, checksums, options and data should all
be identical to the original TCP flow as viewed by
Wireshark with the exceptions noted above.

Figure 16.   Test Network Configuration

1. MySQL Database

This tool will use a MySQL database to store the traits (TCP header fields and data field) of each packet. the program tcpFlowPrepper.py will read in a network trace file, parse the packets one at a time, and send all of the IP packet traits to the database. Once all of the packet traits for a network trace are loaded into a database, the trace file does not need to be read again if the test is to be performed on a different TCP flow contained within the same network trace. Large network trace files could contain many TCP flows, so the user should be able to change the targeted TCP flow by modifying a text file. This allows the test to be performed on different TCP flows from the same network trace to verify the repeatability of the test. For each flow of interest, the database is queried to identify the set of packets specific to that TCP flow and split it into two trace files based on the sending host.

This also enables the tool to perform queries to remove retransmitted packets. Since a retransmission could occur anywhere within the network trace, the entire trace needs to be examined for retransmissions. The database will also be important for expanding this tool to match the MCTSSA requirements. As mentioned previously, the database is also queried to determine the order of sending within the TCP flow and the order is written to a text file (see Figure 15). The query is based on the source and destination IP address and the source and destination port number.

a. Remove TCP Retransmissions

TCP retransmissions should be removed because in the final version of the tool it will be important to test the efficiency of WAN optimizers. If retransmissions were left in, it would artificially skew the efficiency of the optimizer because it would be trying to send duplicate packets to complete the transfer that are not needed. If retransmissions were replayed then the test would not be accurately measuring the efficiency of the data transfer.

2. Two Hosts

Using two hosts in the test models the real world scenario of a file transfer between two hosts. Each host will be transmitting packets based on the order of the original TCP transfer. By doing this, each host will ensure that the packets are replayed in the same relative order as the network trace. A capture of this test should look similar to the original network trace of the flow with IP headers changed to the testing scheme and retransmissions removed.

C. IMPLEMENTATION

The first program uses the Python DPKT module to perform deep packet inspection and determine packet characteristics such as Media Access Control (MAC) address, IP address, protocol types, and data payload. DPKT is a Python module created by Dug Song [17] for reading, writing, and parsing .pcap files. DPKT includes a *reader* and *writer* function that allows for reading and writing .pcap files to determine or alter packet characteristics. Using the dpkt.Reader() function, the original .pcap file is opened and reads one packet from it at a time. DPKT allows the packet to be parsed as it is read into the calling program. The calling program, *tcpFLowPrepper.py*, checks to see if the incoming packet is an IP packet and if so, sends it to a MySQL database. It uses the MySQLdb module to put packet characteristics into an Structured Query Language (SQL) database and perform queries. Once all the packets have been loaded into the database, they will be checked to see if any packets were a retransmission. A MySQL query identifies the flow being targeted. All packets from the targeted TCP flow that were not retransmissions will be written to one of two .pcap files based on source IP address.

Each host will run *orderedReplay.py* to read in the .pcap file and then transmit each packet according to the relative flow order so as to re-create the file transfer between the two hosts. Sequencing of packet transmissions by each host is controlled by maintaining a count of packets sent or received for each sub-flow exchange within the original flow. That is, each host will listen for packets transmitted by the other host and set a flag so that it can check to see if it needs to send the next packet or listen for another packet from the other host. The program will read a text file that contains the order of transmission from the TCP flow of the original network trace file to determine when it is supposed to listen for a packet and when it is supposed to send a packet. When both hosts reach the end of the pertinent text files both programs terminate.

1. PUT ALL IP PACKETS INTO DATABASE

Figure 17 shows the database schema. The schema has two tables, one for IP characteristics and one for TCP characteristics. Some of the packets in the original packet capture may not be IP packets. For the purposes of this experiment, packets that are not IP packets will be discarded. Putting all the IP packets into the database allows the flexibility to identify different TCP flows by changing the query but does not require the original .pcap file to be read in again. To modify which TCP flow to use, the user modifies a text file that provides appropriate parameters to the program.

Table View
For DB PcapParser

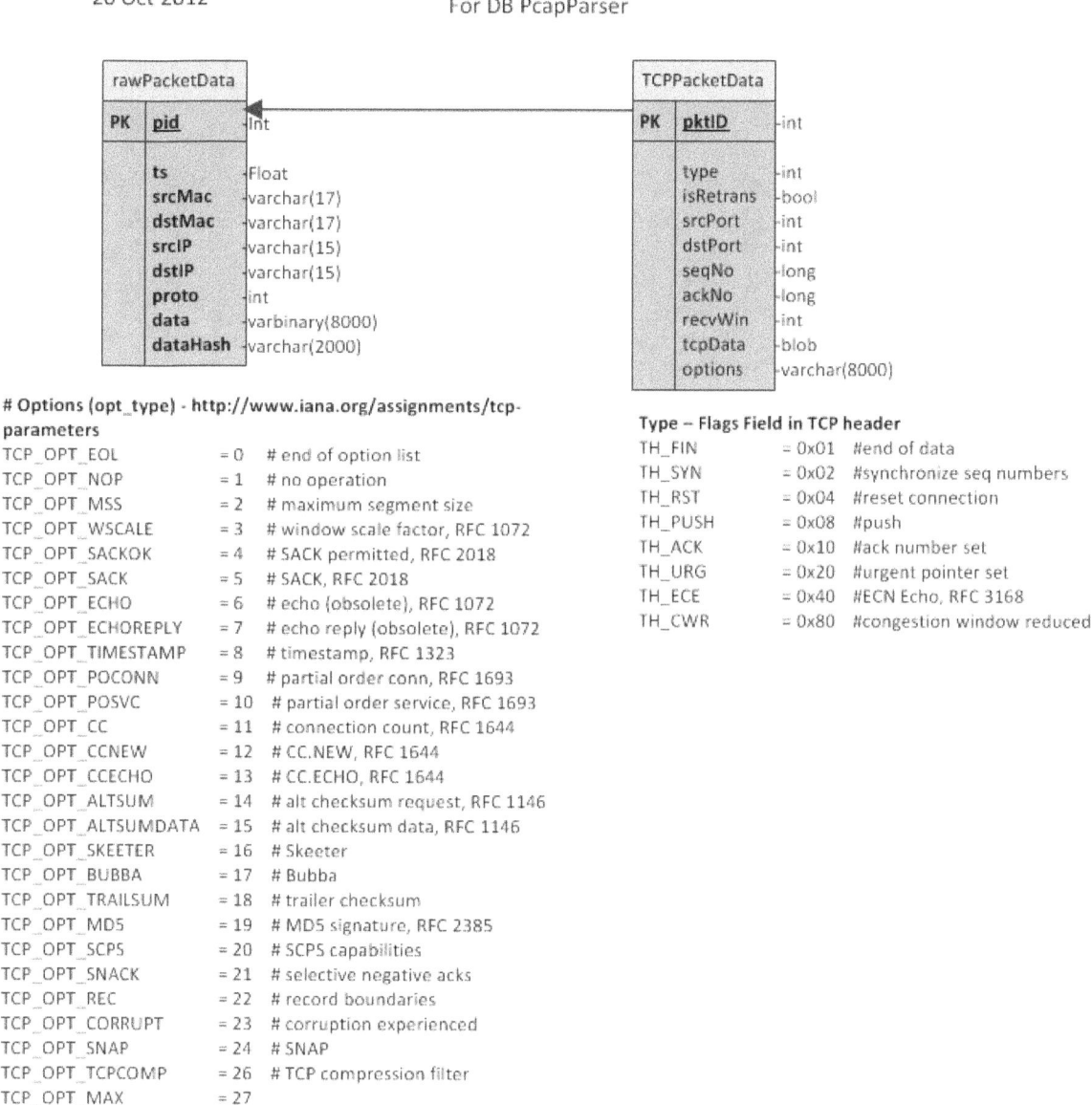

rawPacketData		
PK	**pid**	Int
	ts	Float
	srcMac	varchar(17)
	dstMac	varchar(17)
	srcIP	varchar(15)
	dstIP	varchar(15)
	proto	int
	data	varbinary(8000)
	dataHash	varchar(2000)

TCPPacketData		
PK	**pktID**	int
	type	int
	isRetrans	bool
	srcPort	int
	dstPort	int
	seqNo	long
	ackNo	long
	recvWin	int
	tcpData	blob
	options	varchar(8000)

Options (opt_type) - http://www.iana.org/assignments/tcp-parameters

TCP_OPT_EOL	= 0	# end of option list
TCP_OPT_NOP	= 1	# no operation
TCP_OPT_MSS	= 2	# maximum segment size
TCP_OPT_WSCALE	= 3	# window scale factor, RFC 1072
TCP_OPT_SACKOK	= 4	# SACK permitted, RFC 2018
TCP_OPT_SACK	= 5	# SACK, RFC 2018
TCP_OPT_ECHO	= 6	# echo (obsolete), RFC 1072
TCP_OPT_ECHOREPLY	= 7	# echo reply (obsolete), RFC 1072
TCP_OPT_TIMESTAMP	= 8	# timestamp, RFC 1323
TCP_OPT_POCONN	= 9	# partial order conn, RFC 1693
TCP_OPT_POSVC	= 10	# partial order service, RFC 1693
TCP_OPT_CC	= 11	# connection count, RFC 1644
TCP_OPT_CCNEW	= 12	# CC.NEW, RFC 1644
TCP_OPT_CCECHO	= 13	# CC.ECHO, RFC 1644
TCP_OPT_ALTSUM	= 14	# alt checksum request, RFC 1146
TCP_OPT_ALTSUMDATA	= 15	# alt checksum data, RFC 1146
TCP_OPT_SKEETER	= 16	# Skeeter
TCP_OPT_BUBBA	= 17	# Bubba
TCP_OPT_TRAILSUM	= 18	# trailer checksum
TCP_OPT_MD5	= 19	# MD5 signature, RFC 2385
TCP_OPT_SCPS	= 20	# SCPS capabilities
TCP_OPT_SNACK	= 21	# selective negative acks
TCP_OPT_REC	= 22	# record boundaries
TCP_OPT_CORRUPT	= 23	# corruption experienced
TCP_OPT_SNAP	= 24	# SNAP
TCP_OPT_TCPCOMP	= 26	# TCP compression filter
TCP_OPT_MAX	= 27	

Type – Flags Field in TCP header

TH_FIN	= 0x01	#end of data
TH_SYN	= 0x02	#synchronize seq numbers
TH_RST	= 0x04	#reset connection
TH_PUSH	= 0x08	#push
TH_ACK	= 0x10	#ack number set
TH_URG	= 0x20	#urgent pointer set
TH_ECE	= 0x40	#ECN Echo, RFC 3168
TH_CWR	= 0x80	#congestion window reduced

http://code.google.com/p/dpkt/source/browse/trunk/dpkt/tcp.py?r=42

Figure 17.   Database Schema[After 18]

The key for the IP table is the *pid*. As each packet is read into the database, a counter is incremented. The *pid* is this count and identifies the numerical order in which the packet was received. The *ts* field identifies the timestamp of the packet. This field uses a decimal data type to preserve the precision of the timestamp. The source and destination MAC address and the source and destination IP address are also saved in the IP table. The *proto* field is the protocol within the IP packet, for example, TCP would be 6, and UDP would be 17 [19]. The *data* field is the IP data converted from binary to hex and saved as text. The IP data contains the data encapsulated within the IP datagram. The *datahash* field is an md5 hash of the data within the IP datagram and will be used later to see if the packet data is the same as the IP data in another packet.

The key for the TCP table is the *pktID* and refers to the count of IP packets. Since not all IP packets will contain TCP data, it is necessary to be able to match the IP packet with the TCP packet. The *type* field refers to which TCP flags are set in the incoming packet, i.e. SYN, ACK, FIN, etc. The *isRetrans* field refers to whether or not the packet was marked as a retransmission. The *srcPort* field refers to the TCP source port and *dstPort* field refers to the TCP destination port. The *seqNo* and *ackNo* fields refer to the sequence number and acknowledgment numbers of the incoming packet. The *rxWin* field shows the size of the sending host's receive window. The *tcpData* field is a longtext field that represents the TCP data converted from binary to hex. The *options* field refers to the TCP options set within the incoming TCP packet.

a. *Mark packets that are retransmissions*

The packets are brought into the database and marked as "not a retransmission" *(FALSE*, i.e., *0)*. Once

all the packets are in the database, a sequence of SQL
queries is used from the Python program to see which
packets are retransmissions, and if so identified to change
the *isRetrans* value to *TRUE* for each retransmission. The
sequence of queries determines if the sequence number (SEQ)
is repeated and has data greater than one byte. If the SEQ
is repeated and does not have data then the packet could be
an ACK packet with no data, which would be important to
include (see Figure 18). If the SEQ is repeated and the
data only has 1 byte, then the sending host might be
probing to see if the destination host has a receive window
of 0 (See Figure 19). This would also be an important
packet to include because it indicates flow throttling by
the destination.

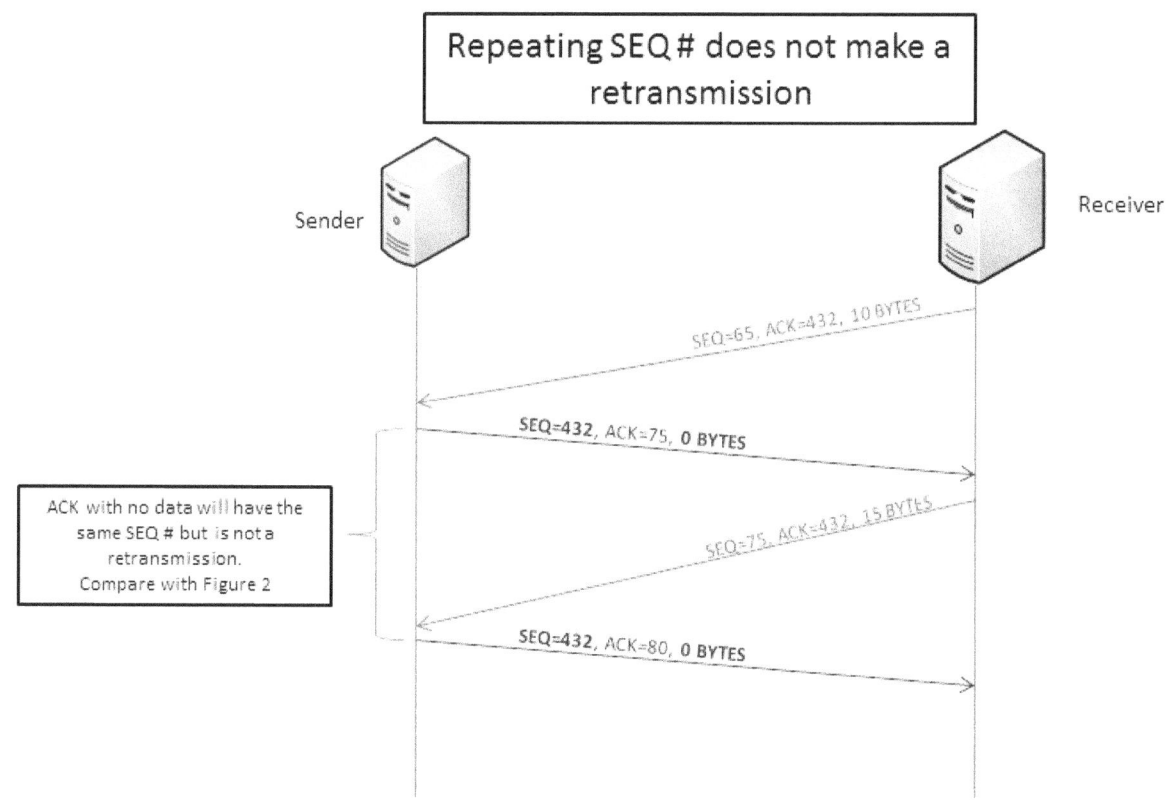

Figure 18. Cases for Retransmission

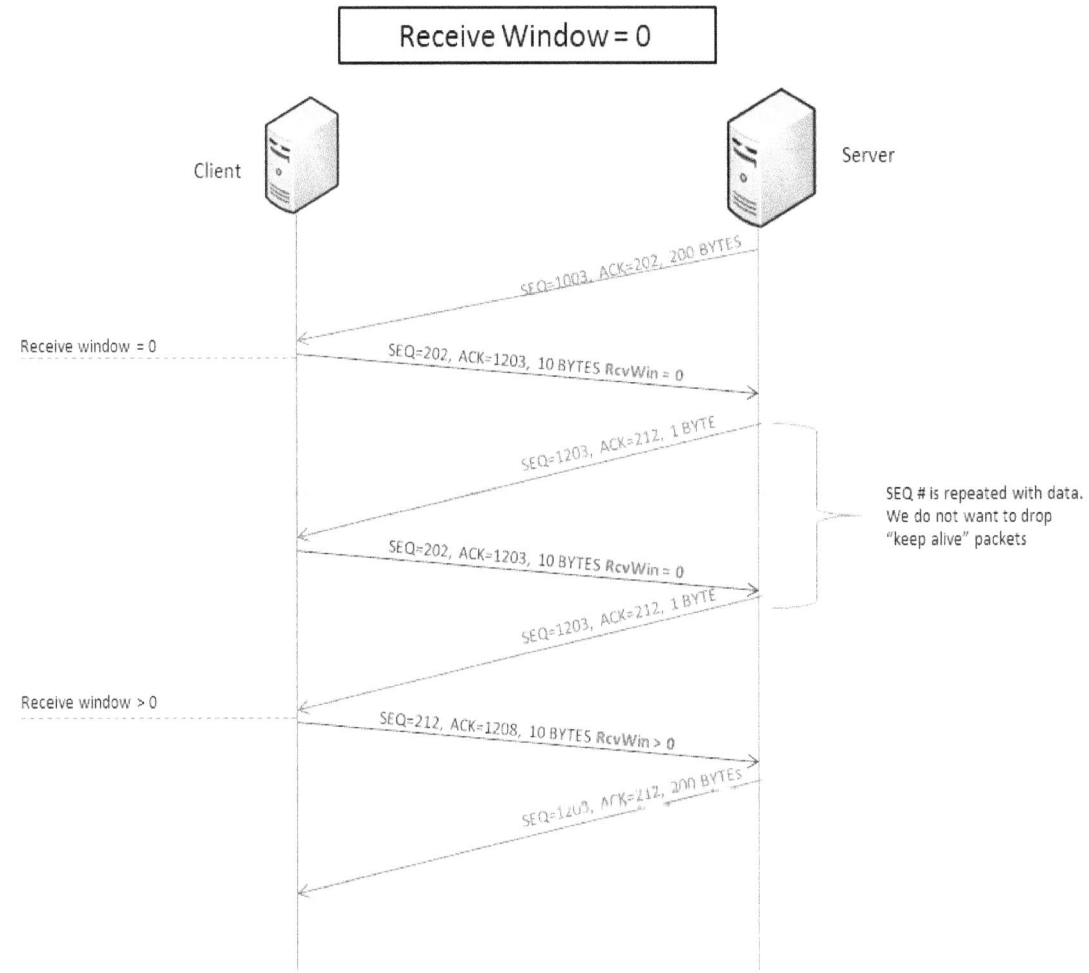

Figure 19.   Cases for Retransmission

The first query gets a list of all the distinct sequence numbers of the TCP packets. A second query uses the list of distinct sequence numbers and checks each sequence number to see if there are any other packets with the same sequence number and if they have a TCP data length greater than one.   A final query updates the *isRetrans* field with a one for packets that are retransmissions.

2. Write the packets to a separate file for sending and receiving hosts

The tool performs an SQL query for the packets sent by each host for the targeted TCP flow. Each query checks to see that the packet belongs to the desired TCP flow, is not a retransmission and that the *srcIP* and *srcPort* fields match (see Figure 16) the sending host in the target TCP flow.

To identify that a packet belongs to a specific TCP flow, five packet characteristics need to be identified. The source and destination IP address, the source and destination port numbers and IP packet type needs to be TCP. These characteristics are read from a text file and used to query the database for packets that match the targeted flow. Packets will be written to a .pcap file based on the sending host. The new .pcap file would then be transferred to the host for sending at a later time.

3. orderedReplay.py

The program, *orderedReplay.py*, reads in the source and destination IP and MAC addresses for the lab setup from a text file. These will be used to change the address scheme from the trace file to the laboratory address scheme. The program will then read the new trace file and put all the packets into a queue. The program will read in the order of transmission one line at a time from a text file and check to see if it should send the first packet in the queue or if it should listen for a packet from the other host. The text file contains the name of the host transmitting and a number representing the number of times the host should transmit before pausing to receive packets (see Figure 15). The program performs the listening using the pylibpcap Python module [20].

Figure 20 shows the decision matrix for *orderedReplay.py*. The program uses a text file (order.txt) containing the order of transmission to determine when to send a packet and when to wait for a signal from *orderedReplay.py* signaling that it received a packet from the other host. Each host is referred to as either host A or host B. If host A is running the program and reads "hosta" in the text file, it knows to send the first packet in the list. If it reads "hostb", it knows to listen for a packet from the other host. This ensures that the sending of packets between host A and host B occur in the proper order. When *orderedReplay.py* reaches the end of order.txt it no longer has any more packets to send or receive and can exit the program. If the program ended when it reached the end of the packet queue it might end while it still had packets to receive from the other host.

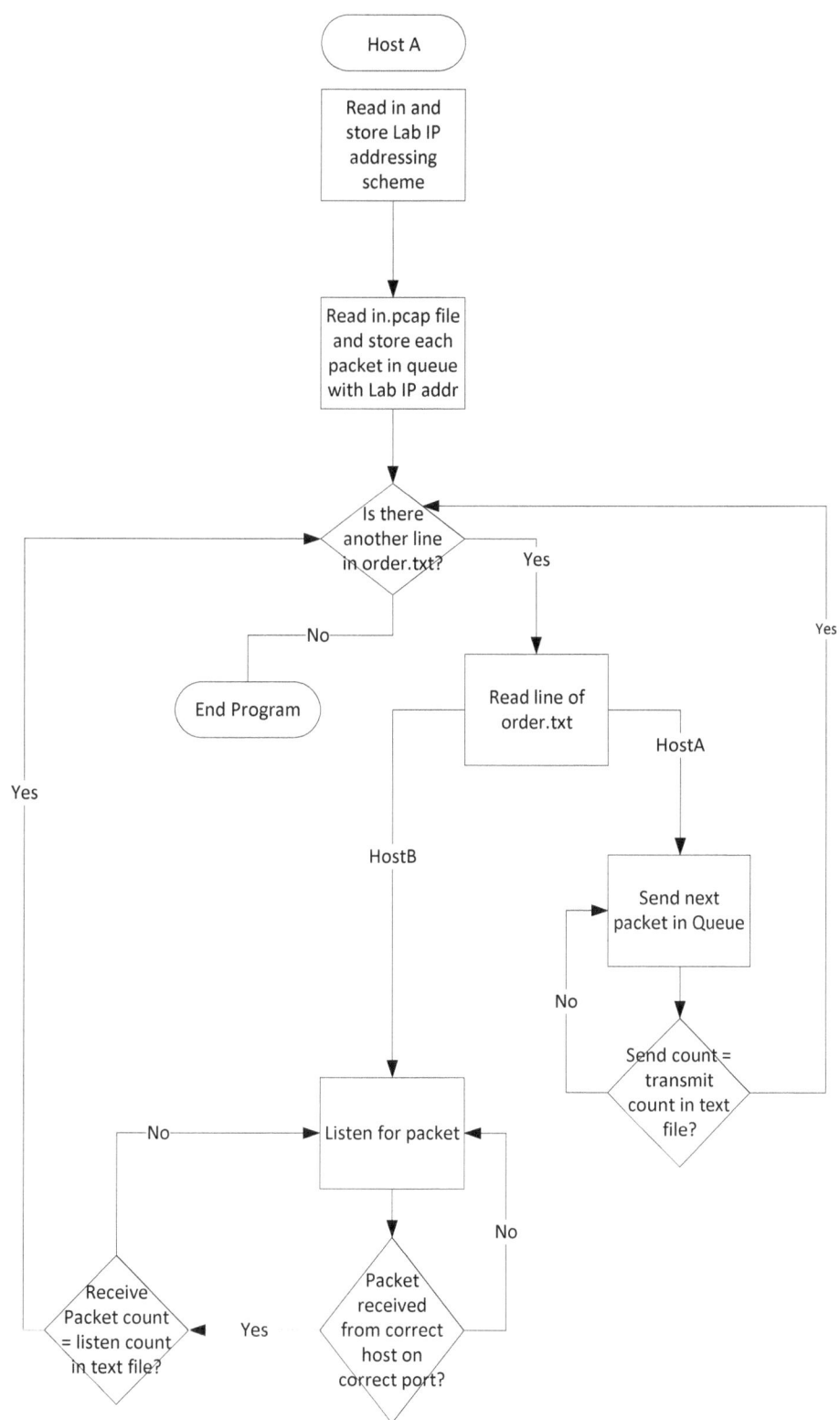

Figure 20.   orderedFlow.py Decision Matrix

The proper sequencing of send and receive actions by each host is critical to the proper execution of the program. Both hosts rely on order.txt to control sequencing of these actions. If the sequence order is interrupted, it will cause a deadlock situation where each host is waiting on the other. If a packet is lost in transit it would disrupt this sequencing and result in a deadlock. The host that would receive that packet would be blocked as it waited for the packet and unable to transition to the send state necessary to perform its next send action. This block state results in the other host also entering a receive state from which it cannot exit, thereby creating the deadlock. Currently, this program does not resolve this issue.

a. Test setup for orderedReplay.py

Both hosts are running Linux Ubuntu 12.04 and are statically configured as shown in Figure 15. Wireless network interface cards (NICs) on both machines are disabled. Each host is configured with an *iptables* rule to prevent the operating system from handling the packets received. The rule applied is: *iptables -A INPUT -s 10.1.1.x -j DROP*. This rule tells the operating system to drop any packets received as input from a host with the source IP address of 10.1.1.x. This is the address range used for the lab set up.

Most applications rely on either TCP or User Datagram Protocol (UDP) to communicate with an associated process on another host. Normally when an application is expecting to receive data on one host from an associated process on another host, it will open a TCP or UDP socket,

60

respectively to listen on depending on the application being used. If the same application also needs to send data and receive to an associated process on another host, it will open a socket for sending. Binding a socket consists of assigning a port and IP address to send to and receive from. The operating system uses the socket to pass data up the stack from the Network Interface Card (NIC) through the datalink and transport layers to the application layer (see Figure 21).

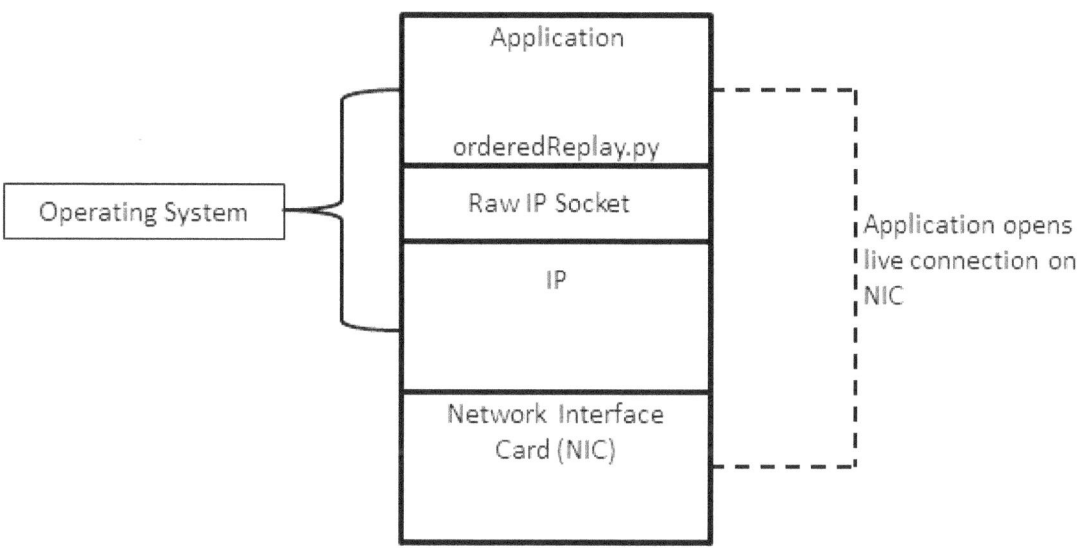

Figure 21. Receiving Socket

The sending host uses the Python *dnet* module to open a raw IP socket and send the packet. This opens a raw IP socket but does not bind it to a port. The receiving host uses the *pcap* module to open a live connection on the NIC and listen for packets (see Figure 21). The receiving host does not open a socket with a binding to a port. Since the socket is not bound to a port on the receiving host, the operating system does not know how to process the packet it receives and responds by sending a reset packet to the originating host. The iptables rule allows the operating system to ignore the incoming packet and the application to use the pcap module to handle the incoming packet, thus preventing the receiver from initiating a reset condition.

If the raw IP socket was bound to a port on the receiving host, the operating system would respond to each packet it received on that port. For example, if the first packet received in the original TCP flow was a SYN, the host operating system would respond with SYN/ACK. However, this would interfere with re-creating the original TCP flow by adding new packets in response to those received.

When a packet arrives, the receiving host filters by destination port number using the Python libpcap module to check that it is addressed to the correct port and coming from the correct host. The receiving host also keeps track of how many packets it has received. This is important because a host may receive several packets before it sends the next, depending on the original TCP flow. Transmitting a packet, however, consists of popping a list of packet traits off of the queue, assigning them to a variable, and sending out the packet. As a result, receiving and parsing the packet to ensure it belongs to the correct flow takes longer than transmitting the packet. To prevent the sending and receiving from getting out of synchronization, a half second delay is added prior to transmitting to ensure that both hosts send and receive in the proper order. If this ordering is not maintained it will result in a deadlock situation as described earlier.

IV. EXPERIMENTAL RESULTS

Data analysis was performed visually using Wireshark, using text print-outs of network trace files, and using a program called *flowStats.py*. The source network trace files were compared to the testing network trace file captured during testing with *orderedReplay.py*. Wireshark provided an easy to use graphical interface to compare packet characteristics. To be more thorough, the source network trace files were also compared to the trace files captured during testing by printing out text files that Wireshark produces for .pcap files. The program *flowStats.py* was used to parse the source network trace and obtain various statistics, including the number of TCP flows per file, the total number of IP packets, the total number of TCP packets, and the average packet size in bytes. It also provided statistics on the targeted TCP flow on the total number of packets, the average packet size, and the number of packets transmitted by each host. To verify the number of retransmissions, Wireshark was compared to the packets marked as retransmitted in the MySQL database for the targeted TCP flow.

A. TEST SETUP

Test 1 was designed to use a short flow to verify that *tcpFlowPrepper.py* and *ordereReplay.py* function correctly. It was also designed to check if retransmissions were removed correctly. Test 1 consisted of replaying only 15 packets, to facilitate manually verifying that all the packets were replayed in the correct order. Test 2 was used to verify that a larger network trace file could be used and that *orderedReplay.py* would function correctly when re-creating a larger TCP flow. Test 3 was designed to verify that a different TCP flow could be targeted from the same network trace by modifying *flow.txt*. Test 4 was designed to use a larger network trace file with a very large TCP flow that dominated the network trace in order to stress test *tcpFlowPrepper.py* and *orderedReplay.py*.

Figure 15 shows the physical setup for the testing conducted. Host A is a Toshiba Satellite L645 laptop with a 2.53 GHz Intel i3 processor and 4GB of RAM. Host B is a Dell Optiplex 745 desktop with a 2.66GHz Intel Core 2 processor and 2 GB of RAM. Host B uses Linux Ubuntu 12.04 LTS as an operating system. All programs were written for Python version 2.7 and use the following Python modules: libpcap 0.6.2, MySQLdb 1.2.3, DPKT 1.6, and dumbnet 1.12-3. To run *orderedReplay.py*, the user on each host computer needs to have root privileges in order to transmit packets.

For each test *tcpFlowPrepper.py* was run on host B. Test 1 and test 3 used *pcap1.pcap* as the source network trace file. This network trace was captured over a period of approximately 30 minutes while connected to the NPS network with several Firefox browser windows open and initiating multiple HTTP sessions. Test 2 used *pcap4.pcap* as the source network trace file. This network trace was captured over approximately 15 minutes while downloading a 5MB file from www.uniblue.com with several other HTTP sessions occurring simultaneoulsy. Test 4 was captured while downloading a 206 MB Microsoft SharePoint 2007 file from the NPS Downloadable Software Library website with other HTTP sessions occurring concurrently to ensure

64

multiple TCP flows were available. During this test it was determined that the source network trace file was too large for tcpFlowPrepper.py to handle and was truncated using Wireshark to 48.5 MB. Table 2 lists which source network file was used for each test. *FlowStats.py* was used to generate the data for Table 2 and Table 3 and was verified visually using Wireshark or by manually inspecting the MySQL database. Table 4 lists how long it to run each program.

Test	File Name	Size	IP Packets	TCP Flows	Average Size
1	pcap1.pcap	4.1 MB	8,878	378	446 B
2	pcap4.pcap	20.8 MB	32,231	842	633 B
3	pcap1.pcap	4.1 MB	8,878	378	446 B
4	Pcap6_2.pcap	48.5 MB	50,503	32	952 B

Table 2.   Network Traces Used in Test

	Packets	Retransmissions	Host A Packets	Host B Packets	Average Size
Test 1	23	8	8	7	125 B
Test 2	5,595	0	1469	4126	1,084 B
Test 3	154	0	68	86	747 B
Test 4	49,817	1	32,851	17,006	962 B

Table 3.   Targeted TCP Flow in Each Test

	tcpFlowPrepper		orderedReplay	flowStats
	Empty Database	Full Database		
Test 1	17 Min	1.5 Min	3 Sec	0.5 sec
Test 2	1 Hr	5.5 Min	1 Hr 15 Min	2 .0 Sec
Test 3	17 Min	1.5 Min	1.25 Min	1.9 Sec
Test 4	16 Hrs	1Hr 25 Min	15 Hrs 7 Min	3.2 Sec

Table 4. Program Performance

For each test, Wireshark was used to search for a TCP flow of interest. Then the source IP address, source port number, destination IP address, and destination port address was manually entered into *flow.txt* to identify the targeted TCP flow for *tcpFlowPrepper.py*. The tcpFlowPrepper.py program produces an output file called *send.txt*. A copy of *send.txt* and the appropriate "host.pcap" file was placed on each host. *HostA.pcap* contains all of the packets that Host A will send to Host B. Likewise, *hostB.pcap* contains all of the packets Host B will send to Host A.

For the program *orderedReplay.py*, *lab_input.txt* was configured with the IP and MAC addresses of Host A and Host B. *Target_flow.txt* was configured with the source and destination IP addresses of Host A and Host B and the source and destination port numbers from the targeted TCP flow. This information was used by *orderedReplay.py* to verify that the host received a packet from the correct TCP flow. Prior to running *orderedReplay.py* the iptables rule must be entered on each host as discussed in chapter 3 (see figure 22).

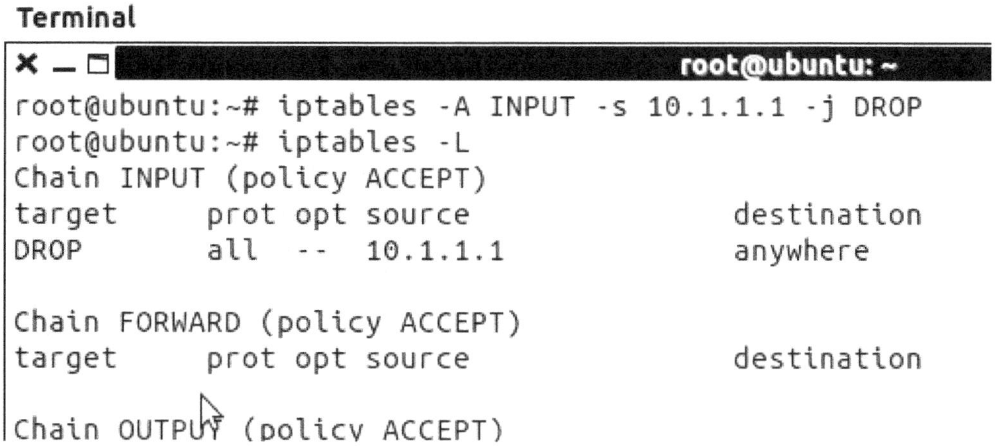

```
Terminal
✖ ─ ☐                                        root@ubuntu: ~
root@ubuntu:~# iptables -A INPUT -s 10.1.1.1 -j DROP
root@ubuntu:~# iptables -L
Chain INPUT (policy ACCEPT)
target     prot opt source              destination
DROP       all  --  10.1.1.1            anywhere

Chain FORWARD (policy ACCEPT)
target     prot opt source              destination

Chain OUTPUT (policy ACCEPT)
```

Figure 22.   iptables Rule

The program *orderedReplay.py* must be modified slightly for Host B. Figure 20 shows the decision tree for Host A to determine if it should send packet(s) or if it should listen for packet(s). For Host B, the program is modified so that if it reads a text line starting with "hostb" from *order.txt*, it sends a number of packets as specified in the second field of the line. Likewise, if it reads "hosta" from *order.txt*, it listens for the specified number of packets.

For each test, *hostB.pcap* contains the SYN packet that is used to start the ordered exchange of packets between the two hosts. For this reason, Host A should be started first so that it is listening when orederedReplay.py is started on Host B and sends the first packet. This ensures that the replaying of packets is started in the correct order to prevent a deadlock situation as discussed in chapter III.

B.   TEST 1 – A SHORT TCP FLOW WITH RETRANSMISSIONS

The TCP flow in test 1 was targeted because it was a short flow that contained retransmissions. This ensures that the flow is short enough to verify by hand that when the flow is re-created, the packets are replayed in the correct order with all, and only, the retransmissions removed. The original TCP flow had 23 packets with eight retransmissions. The program *tcpFlowPrepper.py* ensured the retransmissions were not written to *hostA.pcap* for replay by *orderedReplay.py* and took three seconds to complete. The program *flowStats.py* took one half second to gather statistical data for Test 1 on the source network trace file and the targeted TCP flow. For this test, *tcpFlowPrepper.py*, took 17 minutes to run, including adding packets to the database but only one and half minutes if the packets were already added to the database. Figure 23 shows *pcap1.pcap* with eight retransmissions at the end sent by Host B. Figure 24 shows the screen capture from Host A during the test using *orderedReplay.py* and verifies that the packets were replayed in the correct order with the retransmissions removed. Figure 25 shows the screen capture from Host B during the *orderedReplay.py* test and confirms what is shown in Figure 24. Figures 23 – 25 also show that "Info" block in Wireshark matches between the source network trace and the trace of the replayed TCP flow. Packet length also matches between the source network trace and the replayed network trace with the exception of the last four packets in Figures 24 and 25. Further inspection revealed that the six byte difference was due to an Ethernet trailer. The Ethernet trailer is appended by the Network Interface Card. Depending on the operating system, some operating systems will strip the Ethernet trailer and others will not. Host A does not strip the Ethernet trailer in the previously mentioned frames. However, this does not significantly affect the ability of *tcpFlowPrepper.py* to select a particular TCP flow or *orderedReplay.py* to replay it.

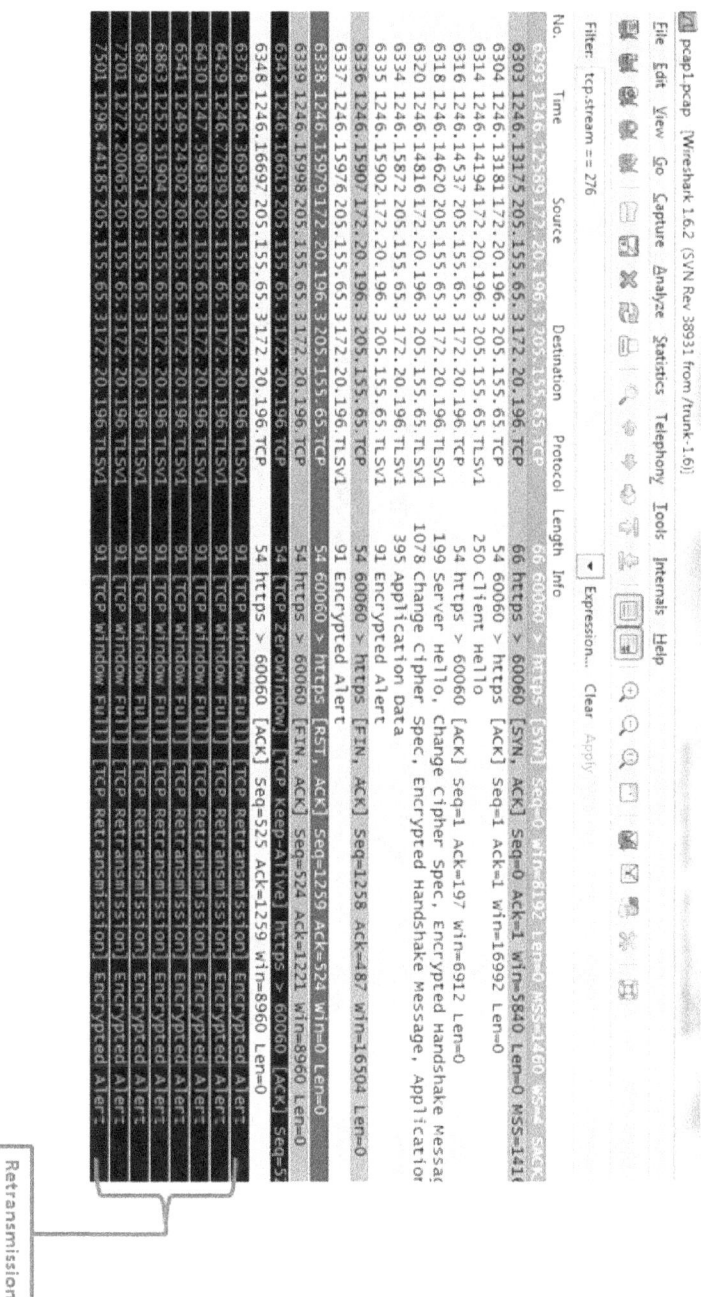

Figure 23. Test 1 Target TCP Flow Screen Capture

test1a.pcap [Wireshark 1.6.2 (SVN Rev 38931 from /trunk-1.6)]

File Edit View Go Capture Analyze Statistics Telephony Tools Internals Help

Filter: tcp ▾ Expression... Clear Apply

No.	Time	Source	Destination	Protocol	Length	Info
7	16.763688	10.1.1.2	10.1.1.1	TCP	66	60060 > https [SYN] Seq=0 Win=8192 Len=0 MSS=1460 WS=4 SACK_PERM=1
8	16.966284	10.1.1.1	10.1.1.2	TCP	66	https > 60060 [SYN, ACK] Seq=0 Ack=1 Win=5840 Len=0 MSS=1416 SACK_PERM=1 WS=12
9	17.168563	10.1.1.2	10.1.1.1	TCP	54	60060 > https [ACK] Seq=1 Ack=1 Win=16992 Len=0
10	17.370131	10.1.1.2	10.1.1.1	TLSv1	250	Client Hello
11	17.572345	10.1.1.1	10.1.1.2	TCP	60	https > 60060 [ACK] Seq=1 Ack=197 Win=6912 Len=0
12	17.774071	10.1.1.1	10.1.1.2	TLSv1	199	Server Hello
13	17.976014	10.1.1.2	10.1.1.1	TLSv1	1078	Change Cipher Spec, Change Cipher Spec, Encrypted Handshake Message
14	18.178352	10.1.1.1	10.1.1.2	TLSv1	395	Change Cipher Spec, Encrypted Handshake Message, Application Data
15	18.380661	10.1.1.1	10.1.1.2	TLSv1	91	Encrypted Alert
16	18.582229	10.1.1.2	10.1.1.1	TCP	54	60060 > https [FIN, ACK] Seq=1258 Ack=487 Win=16504 Len=0
17	18.784582	10.1.1.1	10.1.1.2	TLSv1	91	Encrypted Alert
18	18.986797	10.1.1.1	10.1.1.2	TCP	54	60060 > https [RST, ACK] Seq=1259 Ack=524 Win=0 Len=0
19	19.189404	10.1.1.1	10.1.1.2	TCP	60	https > 60060 [FIN, ACK] Seq=524 Ack=1221 Win=8960 Len=0
20	19.391180	10.1.1.1	10.1.1.2	TCP	60	[TCP ZeroWindow] [TCP Keep-Alive] https > 60060 [ACK] Seq=524 Ack=1258 Win=0 L
21	19.592923	10.1.1.1	10.1.1.2	TCP	60	https > 60060 [ACK] seq=525 Ack=1259 Win=8960 Len=0

Figure 24. Screen Capture from Host A for Test 1

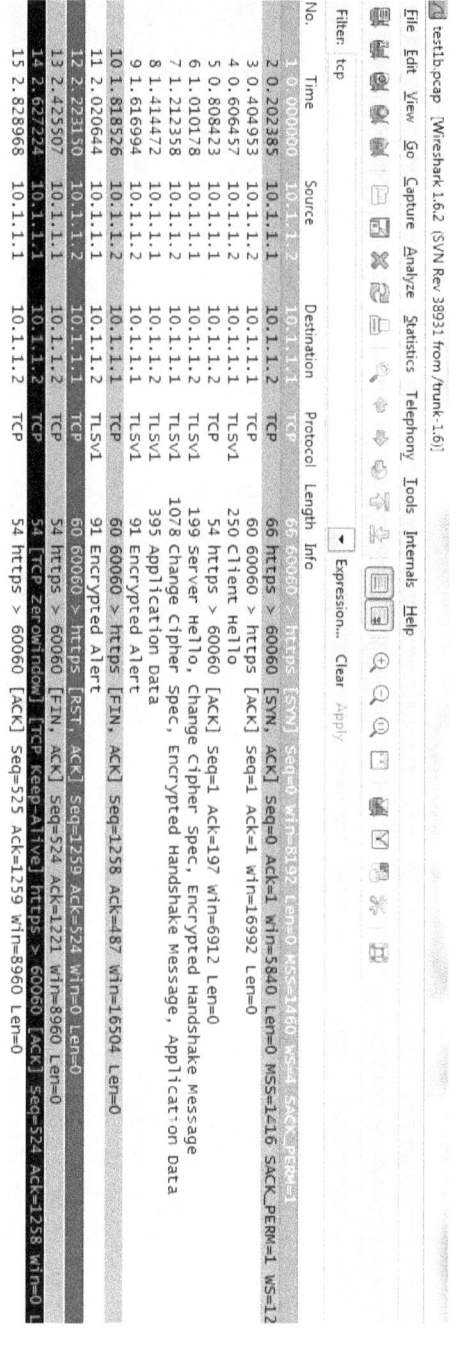

Figure 25. Screen Capture from Host B for Test 1

71

Figure 26 shows a screen capture displaying the details of the SYN packet from *pcap1.pcap* while Figure 27 shows a screen capture of the details of the replayed SYN packet from Host B. These two screen captures verify that the packet was replayed successfully. The IP and MAC addresses were changed, but the port numbers, packet size, TCP header length, options, and window size were unchanged. Figure 28 and Figure 29 show a TCP packet with data in the payload for comparison. The screen captures show that the packets were both the same size, the SEQ and ACK numbers match. They also show the same destination and source port numbers, the same TCP options, and window sizes. For a more detailed comparison of all the packets within the TCP flows, all three network trace files were printed to a text file and compared in order to manually inspect packet details. The check sum is different in Figure 26 than it is in Figure 27 and Figure 28 because the IP and MAC addresses were changed to support the test IP addressing scheme. Figure 28 shows details of a packet with data from the source network trace file. Figure 29 show the same data packet captured on Host A for comparison with Figure 28.

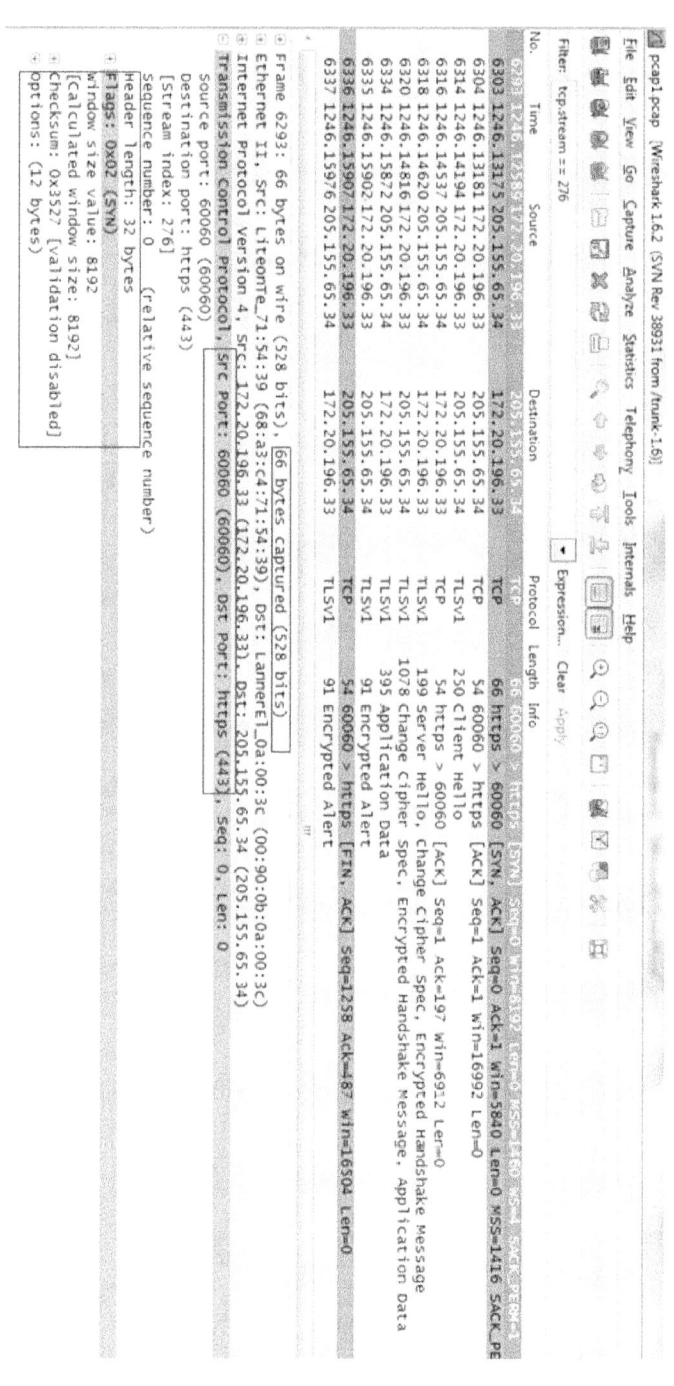

Figure 26. SYN Packet from pcap1.pcap

73

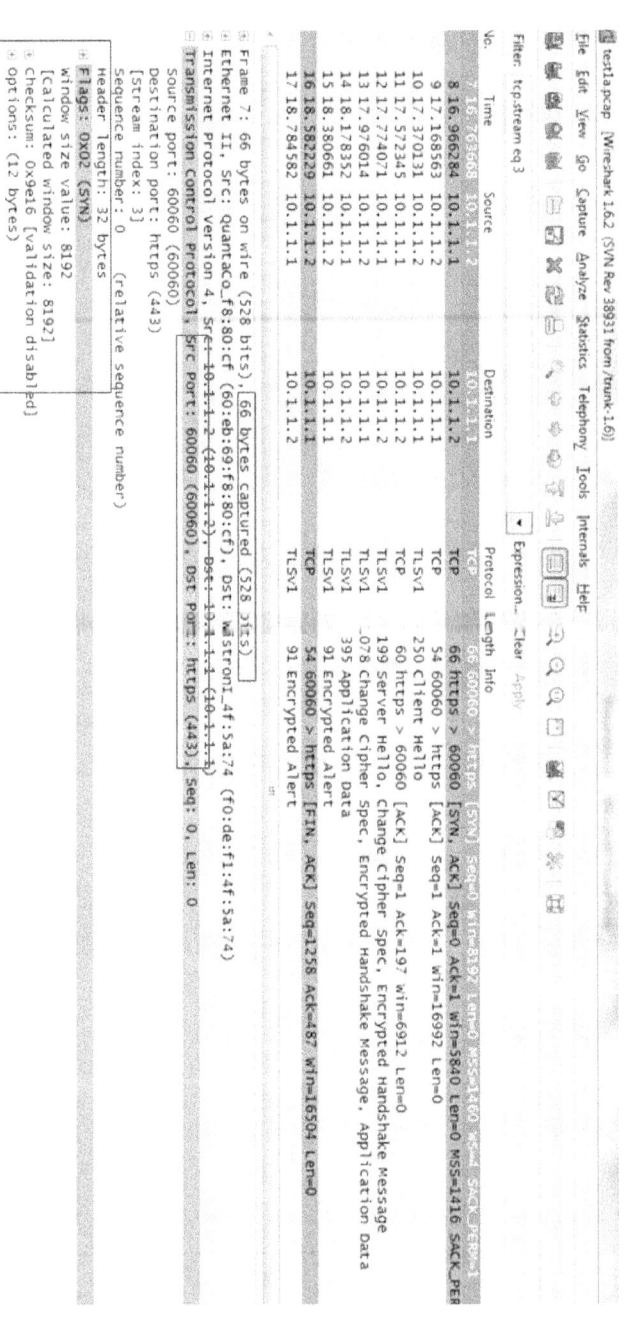

Figure 27. SYN Packet from orderedReplay.py

74

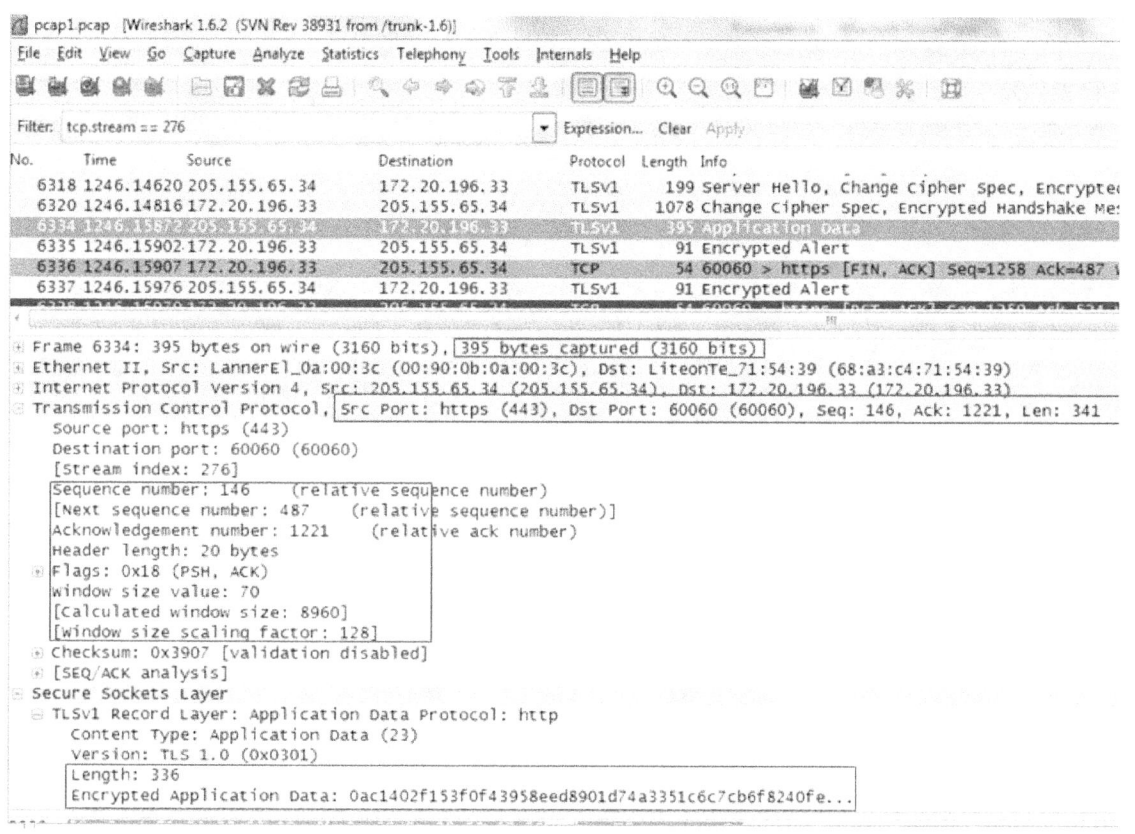

Figure 28.   TCP Packet with Data from pcap1.pcap

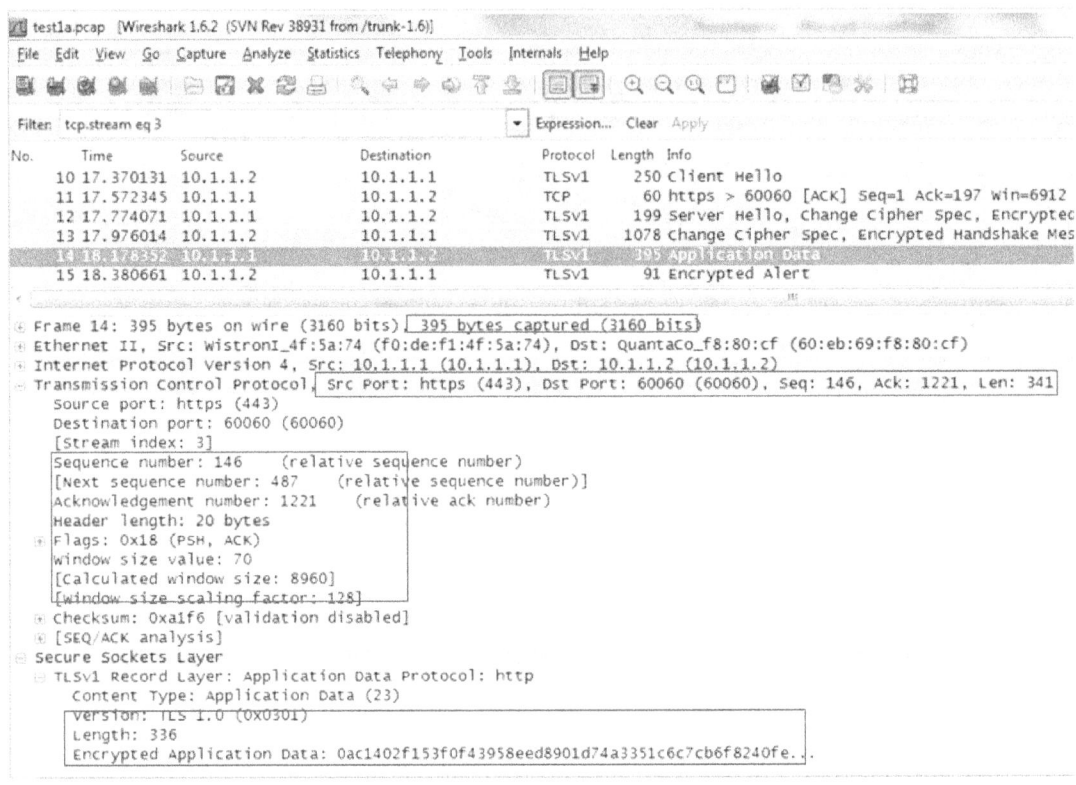

Figure 29. TCP Packet with Data Captured on Host A

For this test there was a 0.2 second delay before transmitting in orderedReplay.py to ensure the packets were transmitted in the correct order without running into a deadlock situation. However, this TCP flow only contained 15 packets so a shorter delay was sufficient. Further testing on longer TCP flows showed 0.5 seconds was required to prevent a deadlock situation. The delay before transmission was required due to the receiving host needing to parse the packet to ensure it is from the correct TCP flow.

C.   TEST 2 — 5MB WEB DOWNLOAD

The TCP flow used in Test 2 was selected to verify that a longer TCP flow could be re-created. The TCP flow used for Test 2 contained 5,595 packets and zero retransmissions. It took one hour to run *tcpFlowPrepper*.py to run with an empty database and five and a half minutes to run with a full database. It took *orderedReplay.py* one hour and 15 minutes to complete sending all the packets. It took two seconds for *flowStats*.py to gather statistical data on the source network trace file and the targeted TCP flow. This was an HTTP download of a 5.6MB driver file that occurred with several other browser windows open to ensure there was cross traffic but that the download would be the dominant TCP flow. Figure 30 shows the source network trace with a filter applied to display the target TCP flow.

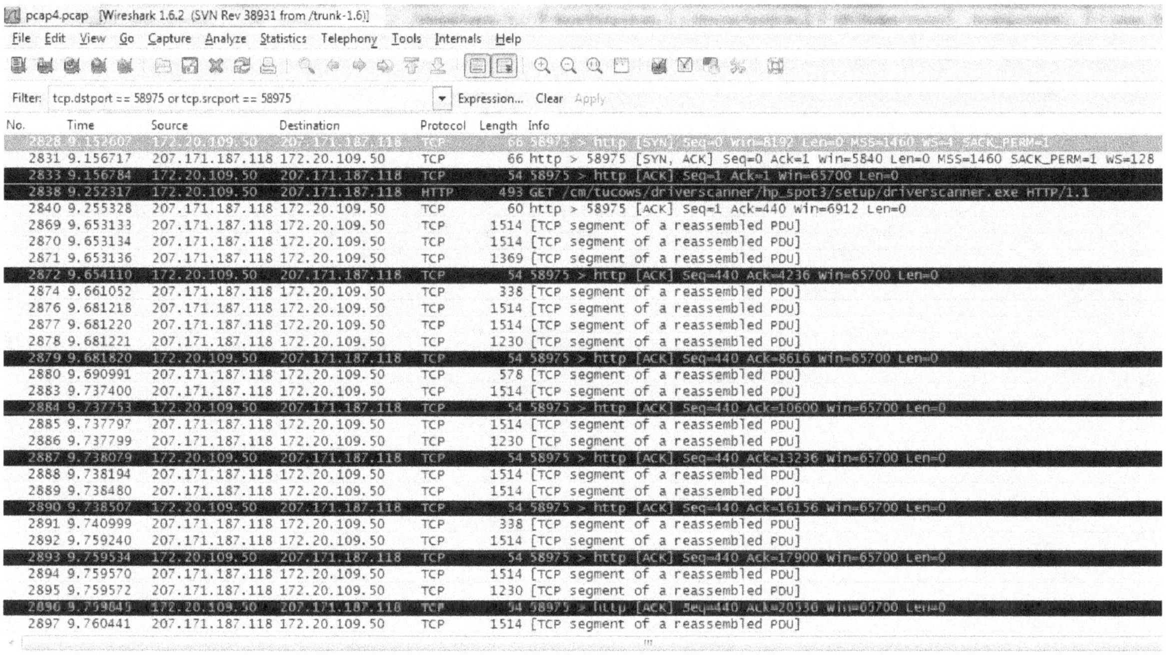

Figure 30. First 30 packets of the Source Network Trace
for Test 2

Tables 2 and 3 show that the average packet size for the entire network trace was 633B and the average packet size of the targeted TCP flow for the download was 1,084B.

77

This TCP flow had 5,595 packets and ran the first time with the 0.5 second delay and only transmitted a little over a 1,000 packets before entering a deadlock state where both hosts were listening for a packet. The delay before packet transmission was increased to 0.7 seconds and each host successfully transmitted all of the packets in order. The delay required before sending a packet caused the test to take approximately 74 minutes to complete. This could cause a problem later when a much larger flow is tested. If 50,000 packets were used, 10 times the number tested here, it would take approximately 13 hours and 40 minutes. If one considers the possibility that the delay between sending may need to be extended again to handle a significantly larger TCP flow during testing the delay could be even longer. As a result, it is recommended that orderedReplay.py be looked at for optimization in future tests to reduce the significant delay during testing and that hosts used for testing have increased processing power and more system memory.

Figure 31 shows the first 30 packets of Test 2 as captured on Host A. Figure 32 shows the first 30 packets of Test 2 captured on Host B. These can be compared with Figure 30 to show the test replayed the packets in the same order as the original network flow. Figure 31 and 32 show there was a TCP filter applied to remove packets not related to the test, like Domain Name Service (DNS) queries and Address Resolution Protocol (ARP) requests. All three trace requests were printed to a text file in order to perform a more detailed examination of all the packets. Figure 31, packet 5, shows the Ethernet trailer is not being removed by the operating system on Host A just like in the first test. Using an outputted text file of the three network trace files, a manual comparison shows that the packet sizes are the same, SEQ and ACK numbers, options, flags and data all match. The checksums are different on the testing hosts than the original network trace because the MAC and IP addresses have been changed to match the testing address scheme. Figure 33 shows the last 30 packets for the targeted TCP flow from the source network trace. Figures 34 and show the network trace captured during Test 2 on Hosts A and B respectively to show the packets were still in order at the end of the test.

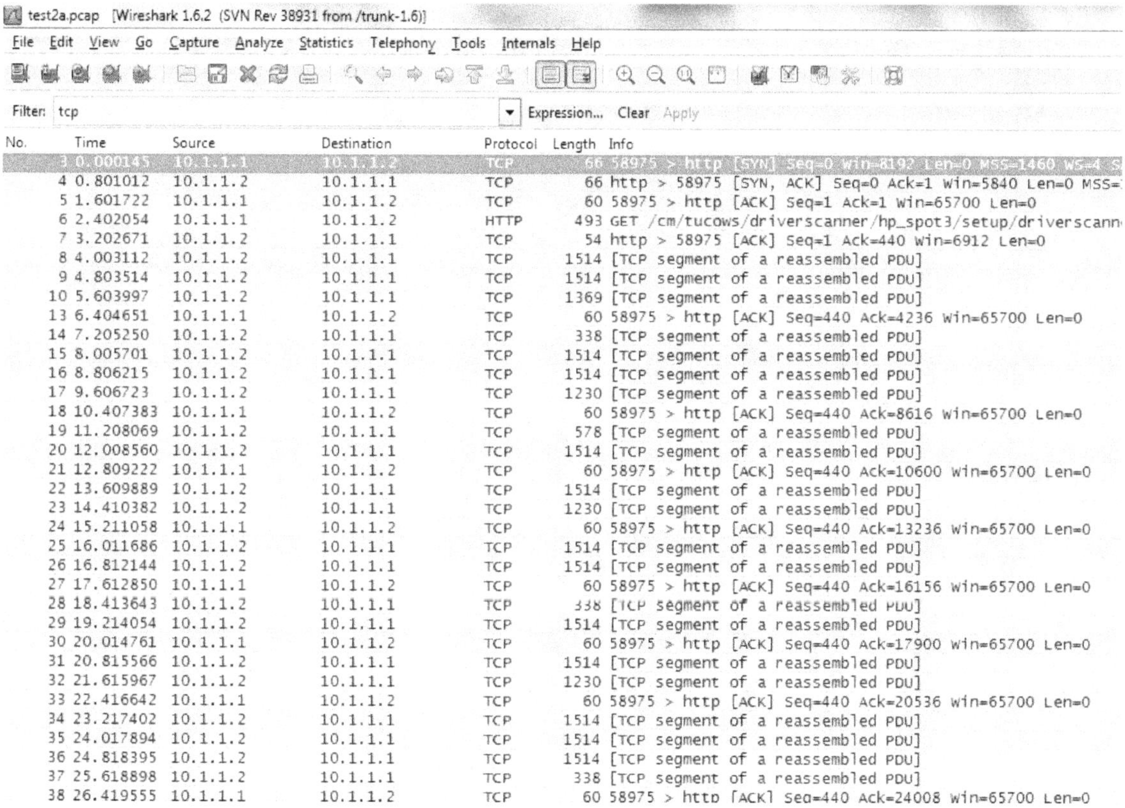
test2a.pcap [Wireshark 1.6.2 (SVN Rev 38931 from /trunk-1.6)]

File Edit View Go Capture Analyze Statistics Telephony Tools Internals Help

Filter: tcp Expression... Clear Apply

No.	Time	Source	Destination	Protocol	Length	Info
3	0.000145	10.1.1.1	10.1.1.2	TCP	66	58975 > http [SYN] Seq=0 Win=8192 Len=0 MSS=1460 WS=4 S
4	0.801012	10.1.1.2	10.1.1.1	TCP	66	http > 58975 [SYN, ACK] Seq=0 Ack=1 Win=5840 Len=0 MSS=
5	1.601722	10.1.1.1	10.1.1.2	TCP	60	58975 > http [ACK] Seq=1 Ack=1 Win=65700 Len=0
6	2.402054	10.1.1.1	10.1.1.2	HTTP	493	GET /cm/tucows/driverscanner/hp_spot3/setup/driverscann
7	3.202671	10.1.1.2	10.1.1.1	TCP	54	http > 58975 [ACK] Seq=1 Ack=440 Win=6912 Len=0
8	4.003112	10.1.1.2	10.1.1.1	TCP	1514	[TCP segment of a reassembled PDU]
9	4.803514	10.1.1.2	10.1.1.1	TCP	1514	[TCP segment of a reassembled PDU]
10	5.603997	10.1.1.2	10.1.1.1	TCP	1369	[TCP segment of a reassembled PDU]
13	6.404651	10.1.1.1	10.1.1.2	TCP	60	58975 > http [ACK] Seq=440 Ack=4236 Win=65700 Len=0
14	7.205250	10.1.1.2	10.1.1.1	TCP	338	[TCP segment of a reassembled PDU]
15	8.005701	10.1.1.2	10.1.1.1	TCP	1514	[TCP segment of a reassembled PDU]
16	8.806215	10.1.1.2	10.1.1.1	TCP	1514	[TCP segment of a reassembled PDU]
17	9.606723	10.1.1.2	10.1.1.1	TCP	1230	[TCP segment of a reassembled PDU]
18	10.407383	10.1.1.1	10.1.1.2	TCP	60	58975 > http [ACK] Seq=440 Ack=8616 Win=65700 Len=0
19	11.208069	10.1.1.2	10.1.1.1	TCP	578	[TCP segment of a reassembled PDU]
20	12.008560	10.1.1.2	10.1.1.1	TCP	1514	[TCP segment of a reassembled PDU]
21	12.809222	10.1.1.1	10.1.1.2	TCP	60	58975 > http [ACK] Seq=440 Ack=10600 Win=65700 Len=0
22	13.609889	10.1.1.2	10.1.1.1	TCP	1514	[TCP segment of a reassembled PDU]
23	14.410382	10.1.1.2	10.1.1.1	TCP	1230	[TCP segment of a reassembled PDU]
24	15.211058	10.1.1.1	10.1.1.2	TCP	60	58975 > http [ACK] Seq=440 Ack=13236 Win=65700 Len=0
25	16.011686	10.1.1.2	10.1.1.1	TCP	1514	[TCP segment of a reassembled PDU]
26	16.812144	10.1.1.2	10.1.1.1	TCP	1514	[TCP segment of a reassembled PDU]
27	17.612850	10.1.1.1	10.1.1.2	TCP	60	58975 > http [ACK] Seq=440 Ack=16156 Win=65700 Len=0
28	18.413643	10.1.1.2	10.1.1.1	TCP	338	[TCP segment of a reassembled PDU]
29	19.214054	10.1.1.2	10.1.1.1	TCP	1514	[TCP segment of a reassembled PDU]
30	20.014761	10.1.1.1	10.1.1.2	TCP	60	58975 > http [ACK] Seq=440 Ack=17900 Win=65700 Len=0
31	20.815566	10.1.1.2	10.1.1.1	TCP	1514	[TCP segment of a reassembled PDU]
32	21.615967	10.1.1.2	10.1.1.1	TCP	1230	[TCP segment of a reassembled PDU]
33	22.416642	10.1.1.1	10.1.1.2	TCP	60	58975 > http [ACK] Seq=440 Ack=20536 Win=65700 Len=0
34	23.217402	10.1.1.2	10.1.1.1	TCP	1514	[TCP segment of a reassembled PDU]
35	24.017894	10.1.1.2	10.1.1.1	TCP	1514	[TCP segment of a reassembled PDU]
36	24.818395	10.1.1.2	10.1.1.1	TCP	1514	[TCP segment of a reassembled PDU]
37	25.618898	10.1.1.2	10.1.1.1	TCP	338	[TCP segment of a reassembled PDU]
38	26.419555	10.1.1.1	10.1.1.2	TCP	60	58975 > http [ACK] Seq=440 Ack=24008 Win=65700 Len=0

Figure 31. First 30 packets Network Trace from Host A During Test 2

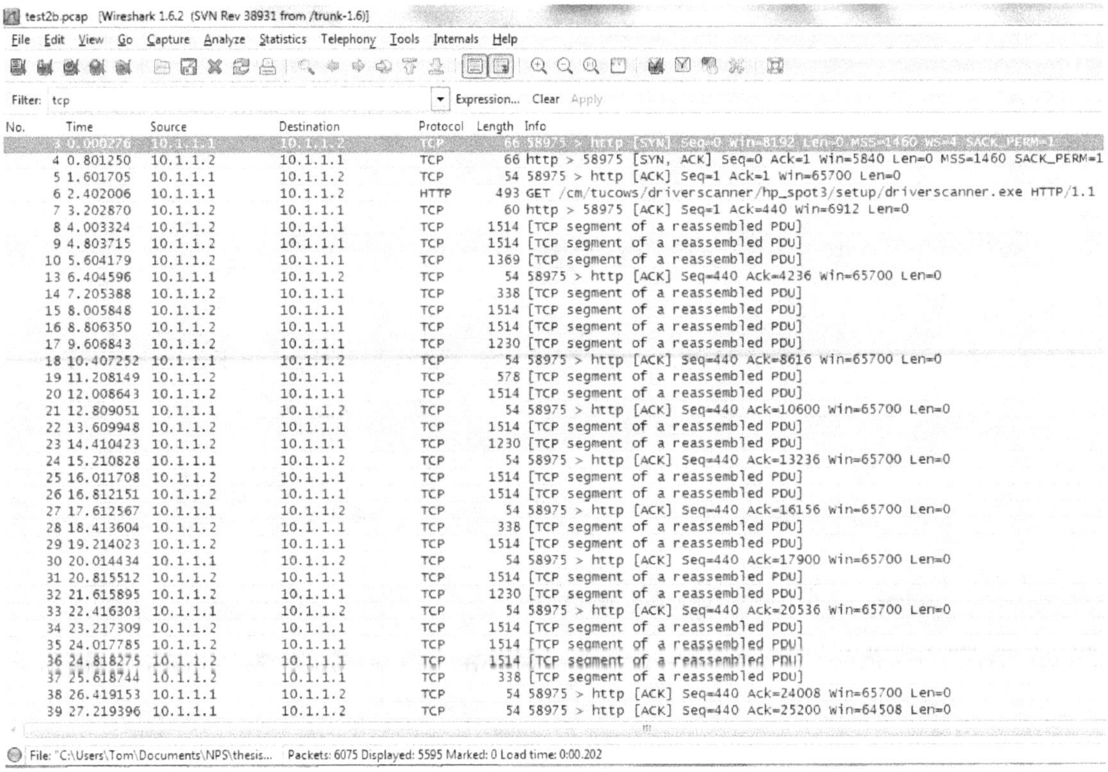

Figure 32.   First 30 packets Network Trace from Host B
During Test 2

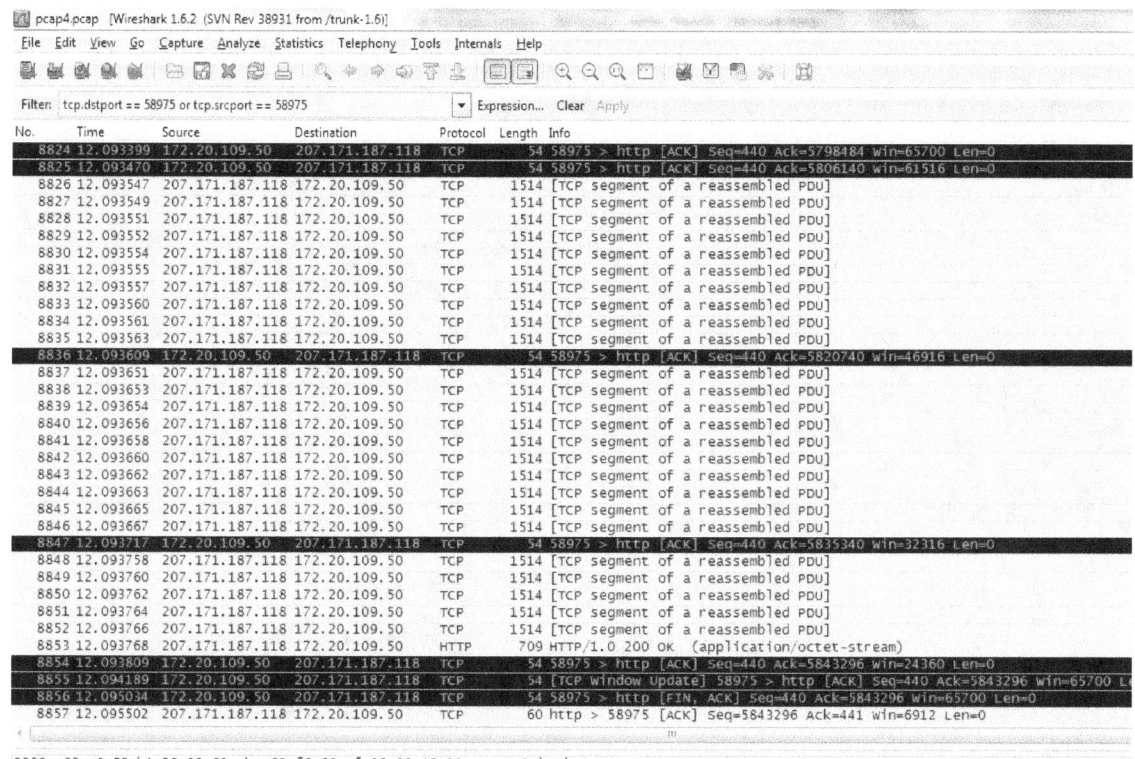

Figure 33. Last 30 packets Source Network Trace for Test 2

```
test2a.pcap  [Wireshark 1.6.2 (SVN Rev 38931 from /trunk-1.6)]

File  Edit  View  Go  Capture  Analyze  Statistics  Telephony  Tools  Internals  Help

Filter: tcp                                        ▼  Expression... Clear  Apply

No.      Time          Source        Destination   Protocol  Length  Info
  6037 4450.87984 10.1.1.1        10.1.1.2         TCP        60 58975 > http [ACK] Seq=440 Ack=5795012 Win=65700 Len=0
  6038 4451.68009 10.1.1.1        10.1.1.2         TCP        60 58975 > http [ACK] Seq=440 Ack=5798484 Win=65700 Len=0
  6039 4452.48034 10.1.1.1        10.1.1.2         TCP        60 58975 > http [ACK] Seq=440 Ack=5806140 Win=61516 Len=0
  6040 4453.28106 10.1.1.2        10.1.1.1         TCP      1514 [TCP segment of a reassembled PDU]
  6041 4454.08146 10.1.1.2        10.1.1.1         TCP      1514 [TCP segment of a reassembled PDU]
  6042 4454.88187 10.1.1.2        10.1.1.1         TCP      1514 [TCP segment of a reassembled PDU]
  6043 4455.68224 10.1.1.2        10.1.1.1         TCP      1514 [TCP segment of a reassembled PDU]
  6044 4456.48264 10.1.1.2        10.1.1.1         TCP      1514 [TCP segment of a reassembled PDU]
  6045 4457.28304 10.1.1.2        10.1.1.1         TCP      1514 [TCP segment of a reassembled PDU]
  6046 4458.08344 10.1.1.2        10.1.1.1         TCP      1514 [TCP segment of a reassembled PDU]
  6047 4458.88384 10.1.1.2        10.1.1.1         TCP      1514 [TCP segment of a reassembled PDU]
  6048 4459.68421 10.1.1.2        10.1.1.1         TCP      1514 [TCP segment of a reassembled PDU]
  6049 4460.48461 10.1.1.2        10.1.1.1         TCP      1514 [TCP segment of a reassembled PDU]
  6050 4461.28529 10.1.1.1        10.1.1.2         TCP        60 58975 > http [ACK] Seq=440 Ack=5820740 Win=46916 Len=0
  6051 4462.08600 10.1.1.2        10.1.1.1         TCP      1514 [TCP segment of a reassembled PDU]
  6052 4462.88651 10.1.1.2        10.1.1.1         TCP      1514 [TCP segment of a reassembled PDU]
  6053 4463.68693 10.1.1.2        10.1.1.1         TCP      1514 [TCP segment of a reassembled PDU]
  6054 4464.48732 10.1.1.2        10.1.1.1         TCP      1514 [TCP segment of a reassembled PDU]
  6055 4465.28781 10.1.1.2        10.1.1.1         TCP      1514 [TCP segment of a reassembled PDU]
  6056 4466.08832 10.1.1.2        10.1.1.1         TCP      1514 [TCP segment of a reassembled PDU]
  6057 4466.88884 10.1.1.2        10.1.1.1         TCP      1514 [TCP segment of a reassembled PDU]
  6058 4467.68932 10.1.1.2        10.1.1.1         TCP      1514 [TCP segment of a reassembled PDU]
  6059 4468.48979 10.1.1.2        10.1.1.1         TCP      1514 [TCP segment of a reassembled PDU]
  6060 4469.29030 10.1.1.2        10.1.1.1         TCP      1514 [TCP segment of a reassembled PDU]
  6061 4470.09100 10.1.1.1        10.1.1.2         TCP        60 58975 > http [ACK] Seq=440 Ack=5835340 Win=32316 Len=0
  6062 4470.89168 10.1.1.2        10.1.1.1         TCP      1514 [TCP segment of a reassembled PDU]
  6063 4471.69208 10.1.1.2        10.1.1.1         TCP      1514 [TCP segment of a reassembled PDU]
  6064 4472.49246 10.1.1.2        10.1.1.1         TCP      1514 [TCP segment of a reassembled PDU]
  6065 4473.29287 10.1.1.2        10.1.1.1         TCP      1514 [TCP segment of a reassembled PDU]
  6068 4474.09326 10.1.1.2        10.1.1.1         TCP      1514 [TCP segment of a reassembled PDU]
  6069 4474.89365 10.1.1.2        10.1.1.1         HTTP      709 HTTP/1.0 200 OK  (application/octet-stream)
  6070 4475.69432 10.1.1.1        10.1.1.2         TCP        60 58975 > http [ACK] Seq=440 Ack=5843296 Win=24360 Len=0
  6071 4476.49456 10.1.1.1        10.1.1.2         TCP        60 [TCP Window Update] 58975 > http [ACK] Seq=440 Ack=5843
  6072 4477.29483 10.1.1.1        10.1.1.2         TCP        60 58975 > http [FIN, ACK] Seq=440 Ack=5843296 Win=65700 I
  6073 4478.09542 10.1.1.2        10.1.1.1         TCP        54 http > 58975 [ACK] Seq=5843296 Ack=441 Win=6912 Len=0
```

Figure 34. Last 30 packets Network Trace from Host A During
Test 2

82

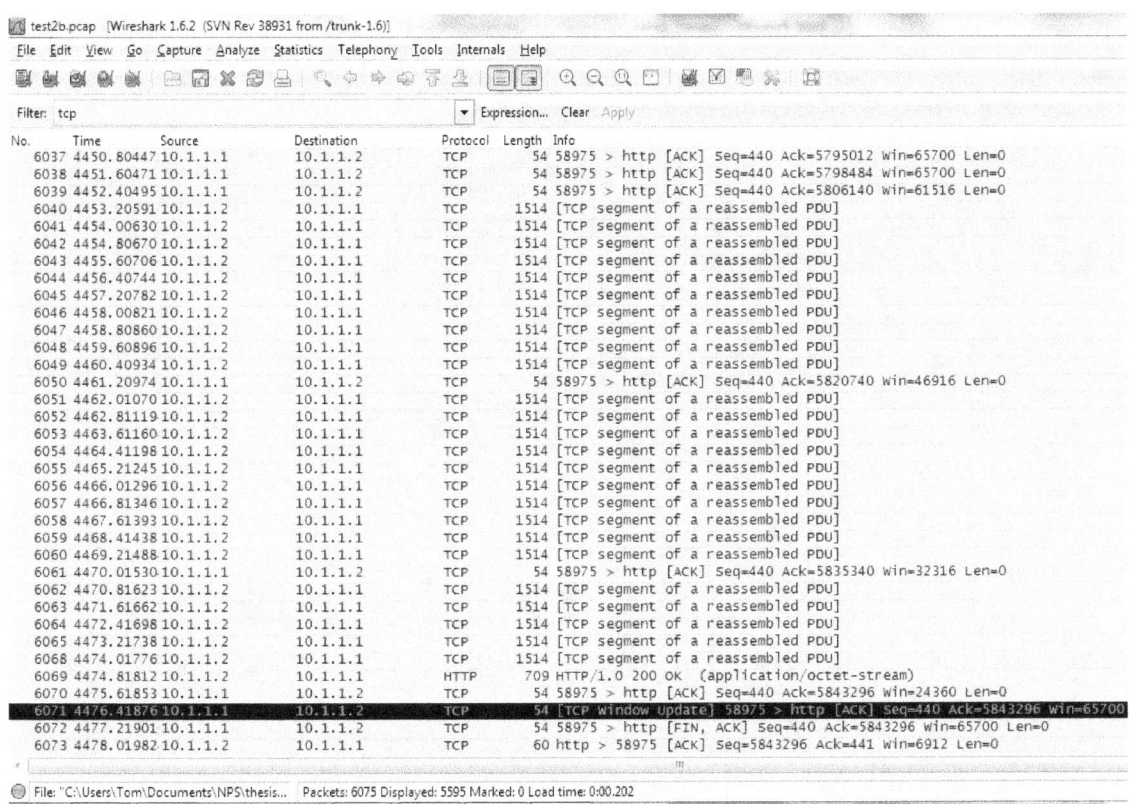

Figure 35.   Last 30 packets Network Trace from Host B During
Test 2

Figure 37 shows detailed information on an HTTP Get message from the source network trace file. Figure 38 and Figure 39 show the same information as captured on Host A and Host B during the test. SEQ, ACK, TCP Flags, window size, window scaling, and HTTP data all match between the source network trace file and the data captured on Host A and Host B during the test.

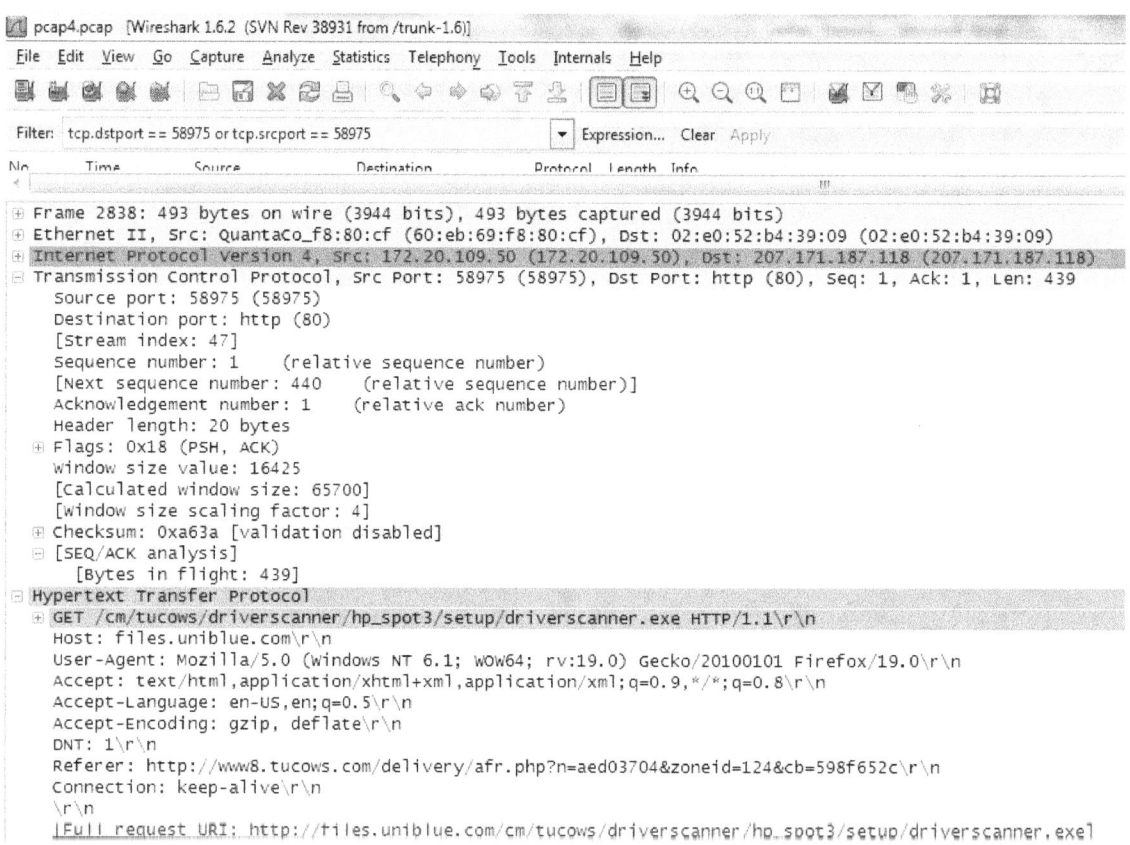

Figure 36.   Test 2 HTTP Get Message from Source Network
Trace

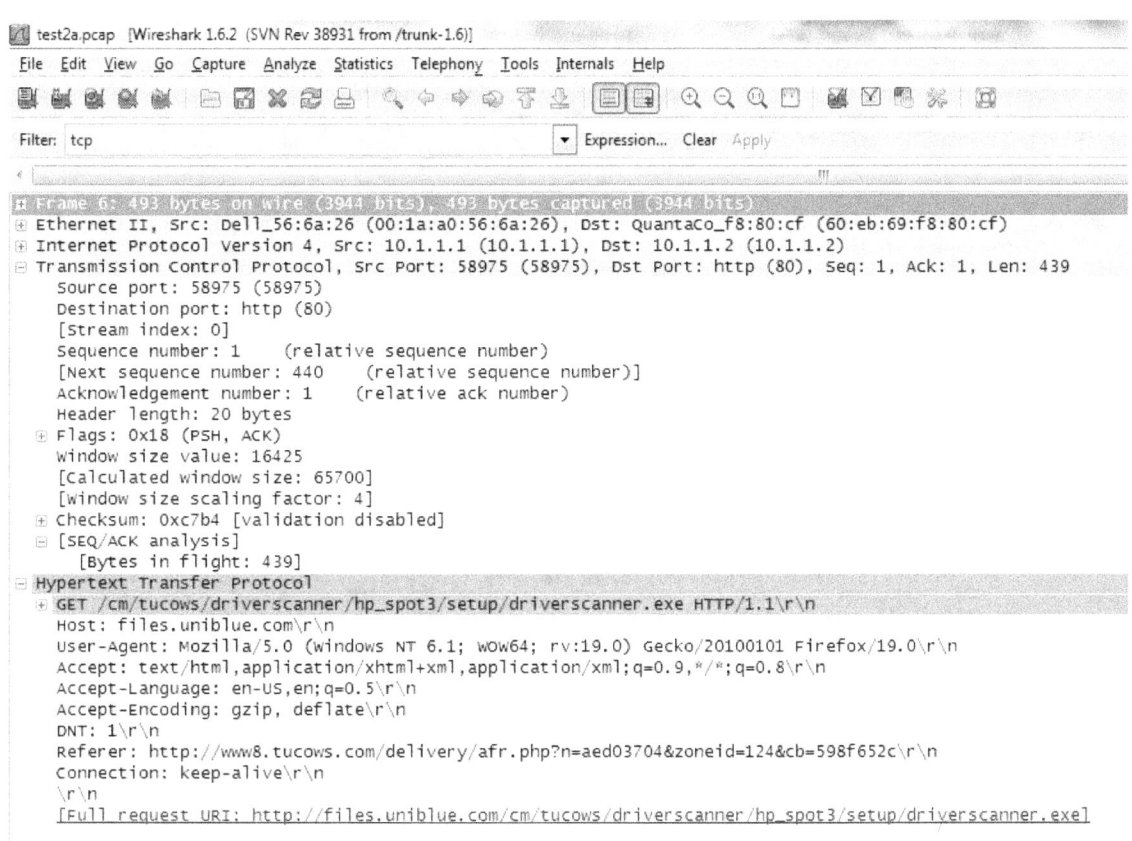

Figure 37. Test 2 HTTP Get Message from Host A

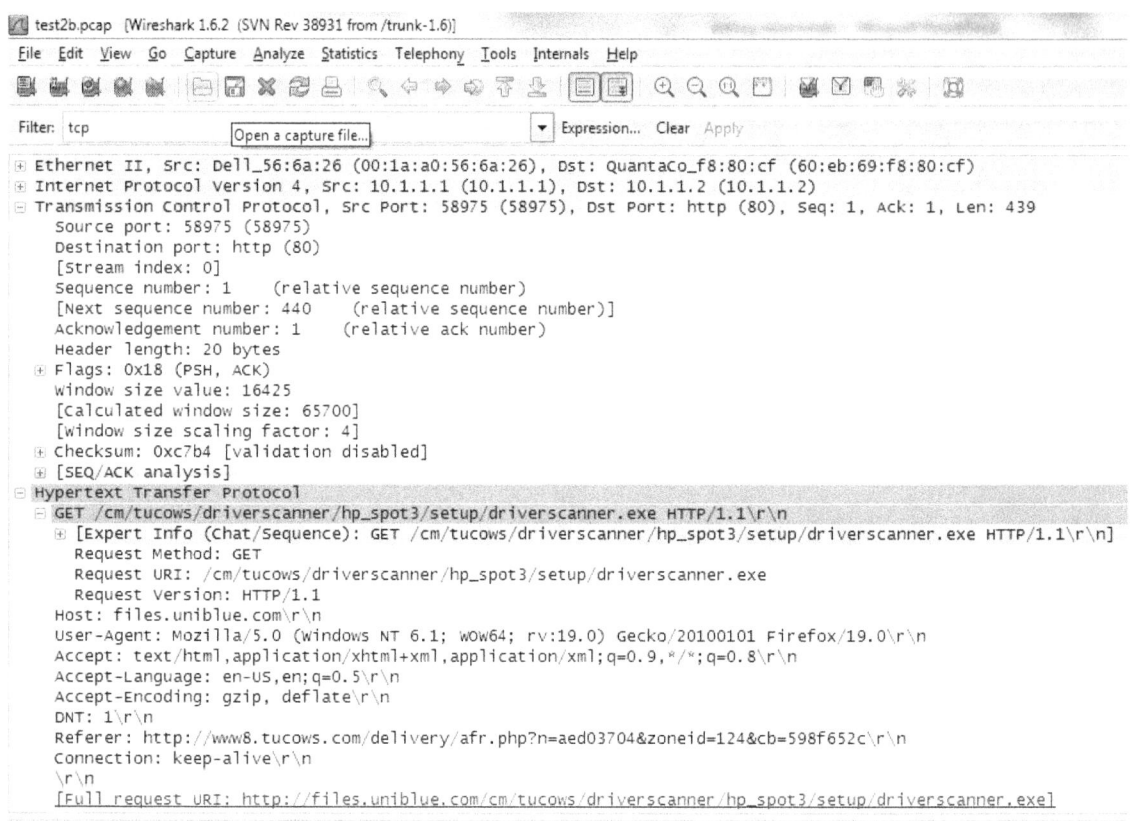

Figure 38. Test 2 HTTP Get Message from Host B

D.   TEST 3 — SAME NETWORK TRACE, DIFFERENT TCP FLOW

Test 3 uses the same source network trace file as Test 1 but targeted a different, longer TCP flow. This test was designed to increase stress on orderedReplay.py from Test 1. Table 3 shows that this TCP flow had 154 packets with zero retransmissions and an average of 747 bytes per packet. The TCP flow tested in Test 1 had an average of 125 bytes per packet and the average packet size for the source network file was 425 bytes. This shows that the targeted TCP flow was a dominant flow within the source network trace. There was a 0.5 second delay before sending packets during the test to prevent the deadlock situation discussed in Chapter III.

Table 4 shows that it took *tcpFLowPrepper.py* 17 minutes to complete with an empty database and one and a half minutes with a full database. It is worth noting that the performance data of Test 1 and Test 3 are identical with an empty database. The test performance should be very similar because they are both using the same source network trace file and entering the same number packets into the database. Test 1 and Test 3 also performed very similarly with a full database because it does not take very long to write the packets and neither test had a very large TCP flow. Test 3 took one minute 15 seconds to complete sending all the packets using *orderedReplay.py* which is significantly longer than Test 1. The program *flowStats.py* took 1.9 seconds to complete gathering statistical data on the source network trace file and the targeted TCP flow.

Figure 40 shows the first 35 packets in the targeted TCP flow from the source network trace file. Figure 41 shows the first 35 packets captured during the test on Host A. Figure 42 shows the first 35 packets captured during the test on Host B. This demonstrates that the captures on Host A and Host B show the same ordering of packets as the targeted TCP flow from the source network file. Many packet characteristics can also be verified to be the same between the targeted TCP flow in the source network trace file and the trace files captured on Host A and Host B during the test. All three trace files, *pcap1.pcap, test3a.pcap, and test3b.pcap*, were printed to a text file and manually inspected to verify that the packet characteristics matched between the source network file and the trace files captured on Host A and Host B during the test. Figure 43 shows the last 35 packets from the

targeted TCP flow in the source network trace file. Figure
44 and Figure 45 show the last 35 packets captured during
the test from Host A and B respectively. The screen
captures show the reader that the packets were still in
order upon completion of the test.

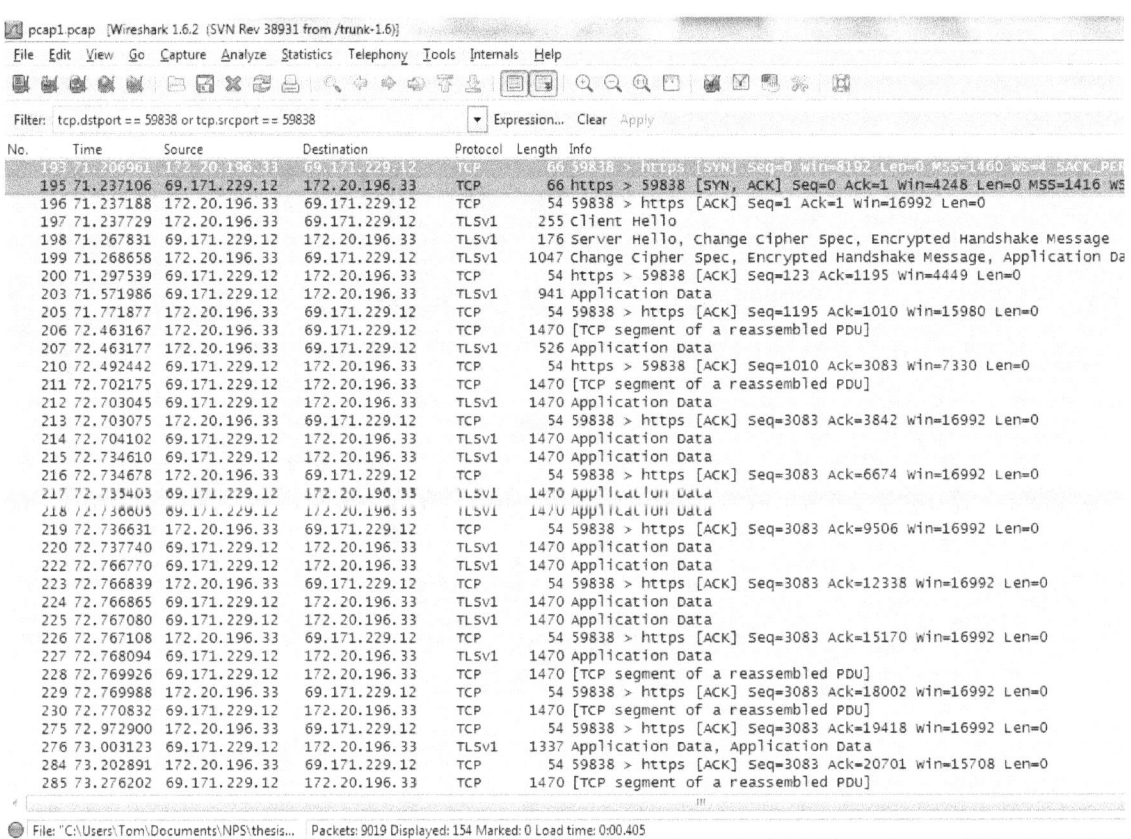

Figure 39.   First 30 packets from Source Network Trace
During Test 2

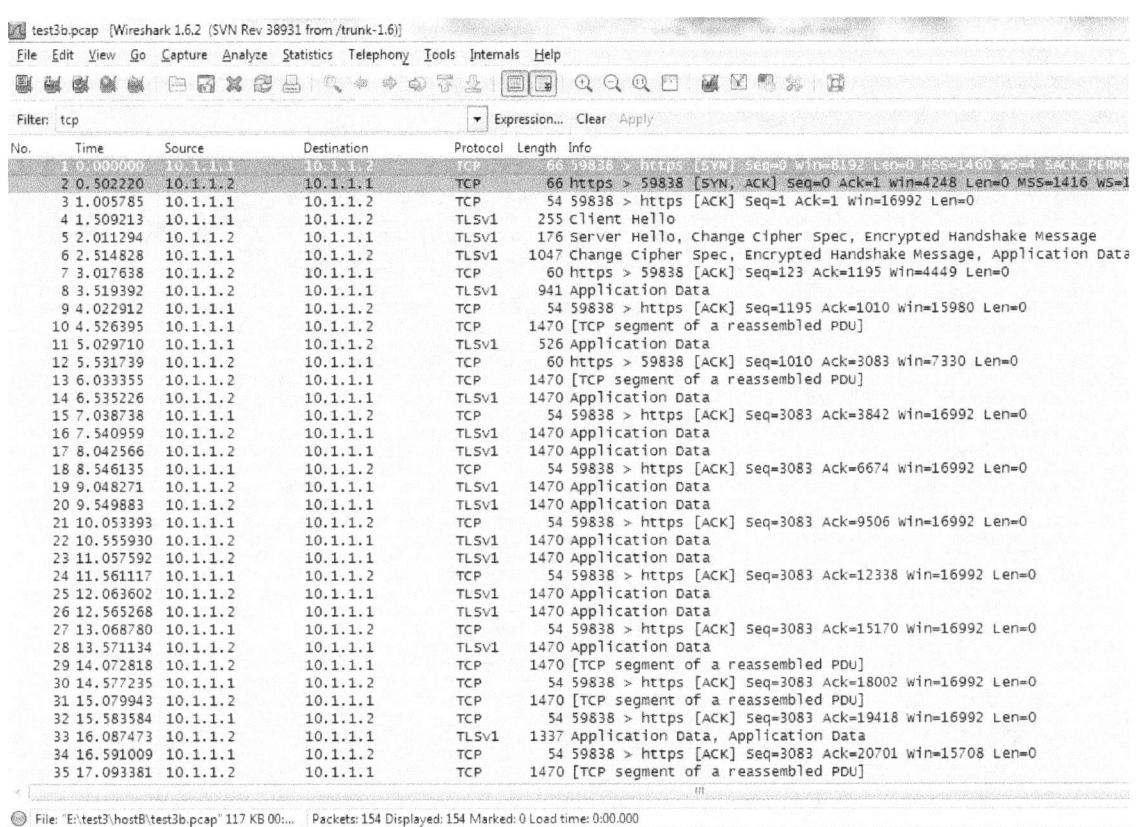

Figure 40.   First 30 packets Captured on Host A During
Test 2

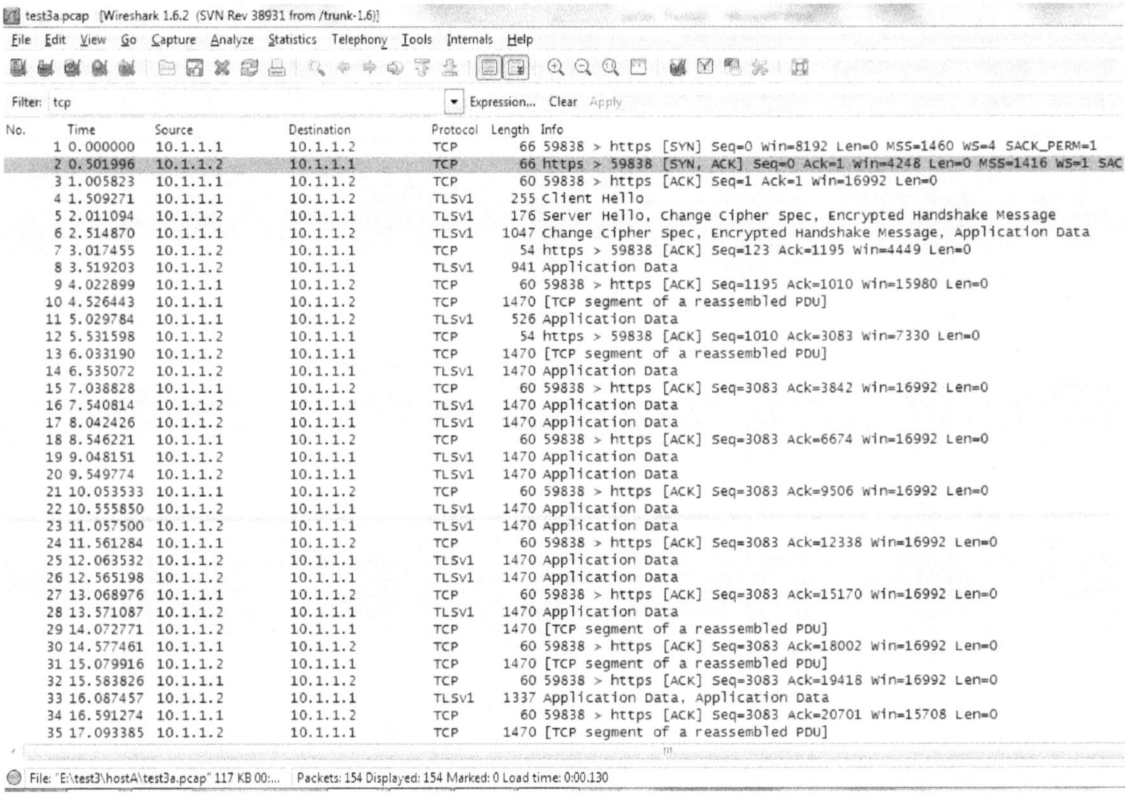

Figure 41.   First 30 packets Captured on Host B During Test 2

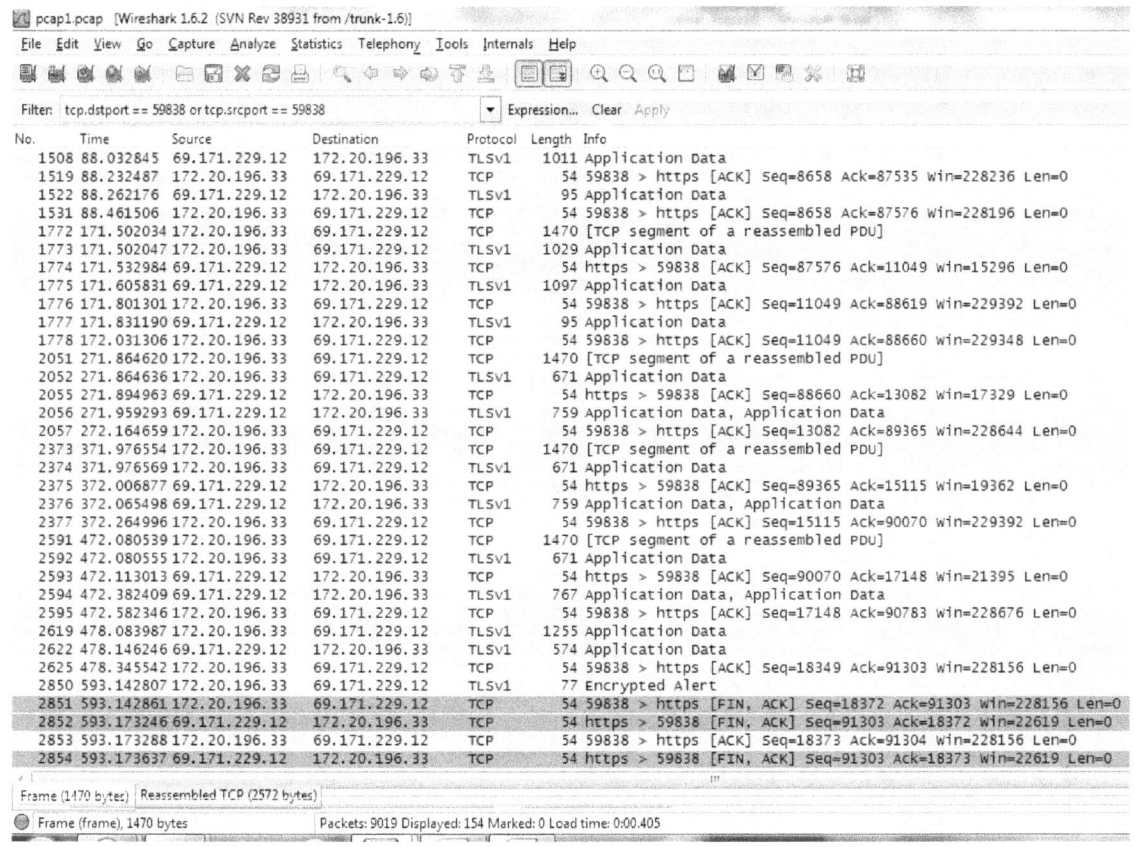

Figure 42.   First 30 packets From Network Trace File

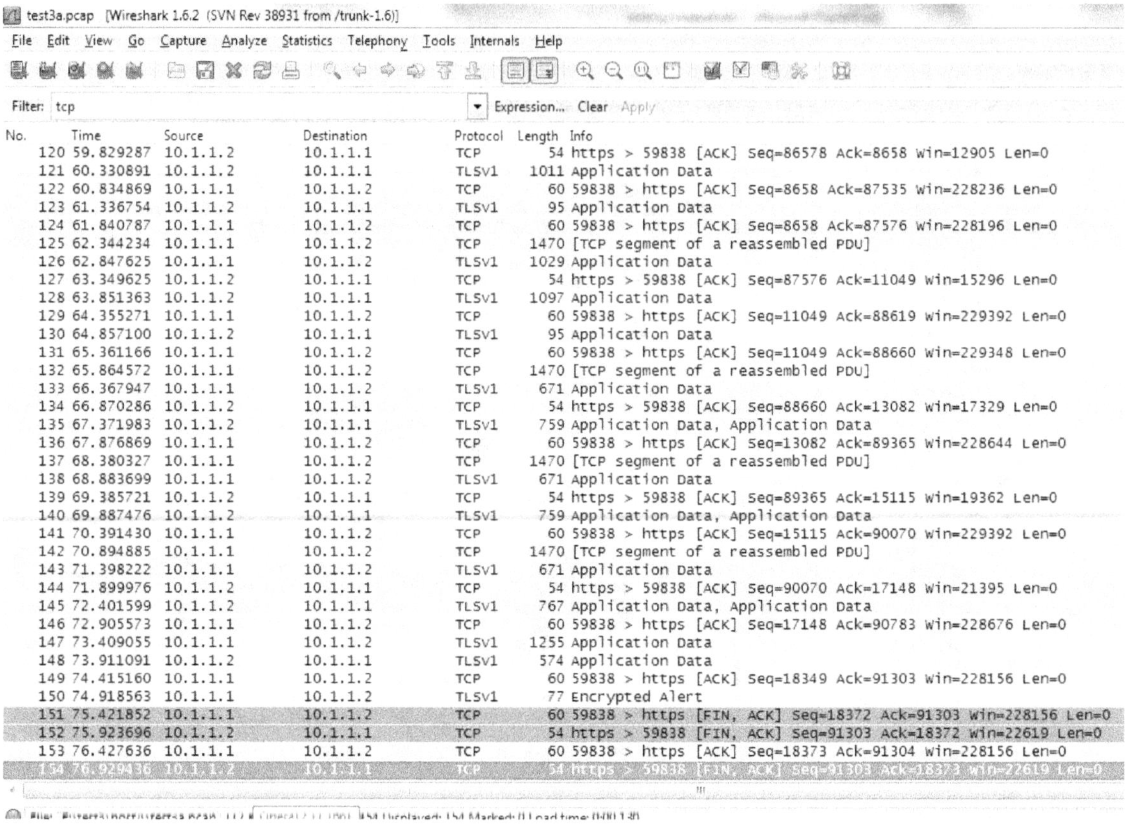

No.	Time	Source	Destination	Protocol	Length	Info
120	59.829287	10.1.1.2	10.1.1.1	TCP	54	https > 59838 [ACK] Seq=86578 Ack=8658 Win=12905 Len=0
121	60.330891	10.1.1.2	10.1.1.1	TLSv1	1011	Application Data
122	60.834869	10.1.1.1	10.1.1.2	TCP	60	59838 > https [ACK] Seq=8658 Ack=87535 Win=228236 Len=0
123	61.336754	10.1.1.2	10.1.1.1	TLSv1	95	Application Data
124	61.840787	10.1.1.1	10.1.1.2	TCP	60	59838 > https [ACK] Seq=8658 Ack=87576 Win=228196 Len=0
125	62.344234	10.1.1.1	10.1.1.2	TCP	1470	[TCP segment of a reassembled PDU]
126	62.847625	10.1.1.1	10.1.1.2	TLSv1	1029	Application Data
127	63.349625	10.1.1.2	10.1.1.1	TCP	54	https > 59838 [ACK] Seq=87576 Ack=11049 Win=15296 Len=0
128	63.851363	10.1.1.2	10.1.1.1	TLSv1	1097	Application Data
129	64.355271	10.1.1.1	10.1.1.2	TCP	60	59838 > https [ACK] Seq=11049 Ack=88619 Win=229392 Len=0
130	64.857100	10.1.1.2	10.1.1.1	TLSv1	95	Application Data
131	65.361166	10.1.1.1	10.1.1.2	TCP	60	59838 > https [ACK] Seq=11049 Ack=88660 Win=229348 Len=0
132	65.864572	10.1.1.1	10.1.1.2	TCP	1470	[TCP segment of a reassembled PDU]
133	66.367947	10.1.1.1	10.1.1.2	TLSv1	671	Application Data
134	66.870286	10.1.1.2	10.1.1.1	TCP	54	https > 59838 [ACK] Seq=88660 Ack=13082 Win=17329 Len=0
135	67.371983	10.1.1.2	10.1.1.1	TLSv1	759	Application Data, Application Data
136	67.876869	10.1.1.1	10.1.1.2	TCP	60	59838 > https [ACK] Seq=13082 Ack=89365 Win=228644 Len=0
137	68.380327	10.1.1.1	10.1.1.2	TCP	1470	[TCP segment of a reassembled PDU]
138	68.883699	10.1.1.1	10.1.1.2	TLSv1	671	Application Data
139	69.385721	10.1.1.2	10.1.1.1	TCP	54	https > 59838 [ACK] Seq=89365 Ack=15115 Win=19362 Len=0
140	69.887476	10.1.1.2	10.1.1.1	TLSv1	759	Application Data, Application Data
141	70.391430	10.1.1.1	10.1.1.2	TCP	60	59838 > https [ACK] Seq=15115 Ack=90070 Win=229392 Len=0
142	70.894885	10.1.1.1	10.1.1.2	TCP	1470	[TCP segment of a reassembled PDU]
143	71.398222	10.1.1.1	10.1.1.2	TLSv1	671	Application Data
144	71.899976	10.1.1.2	10.1.1.1	TCP	54	https > 59838 [ACK] Seq=90070 Ack=17148 Win=21395 Len=0
145	72.401599	10.1.1.2	10.1.1.1	TLSv1	767	Application Data, Application Data
146	72.905573	10.1.1.1	10.1.1.2	TCP	60	59838 > https [ACK] Seq=17148 Ack=90783 Win=228676 Len=0
147	73.409055	10.1.1.1	10.1.1.2	TLSv1	1255	Application Data
148	73.911091	10.1.1.2	10.1.1.1	TLSv1	574	Application Data
149	74.415160	10.1.1.1	10.1.1.2	TCP	60	59838 > https [ACK] Seq=18349 Ack=91303 Win=228156 Len=0
150	74.918563	10.1.1.1	10.1.1.2	TLSv1	77	Encrypted Alert
151	75.421852	10.1.1.1	10.1.1.2	TCP	60	59838 > https [FIN, ACK] Seq=18372 Ack=91303 Win=228156 Len=0
152	75.923696	10.1.1.2	10.1.1.1	TCP	54	https > 59838 [FIN, ACK] Seq=91303 Ack=18372 Win=22619 Len=0
153	76.427636	10.1.1.1	10.1.1.2	TCP	60	59838 > https [ACK] Seq=18373 Ack=91304 Win=228156 Len=0
154	76.929436	10.1.1.2	10.1.1.1	TCP	54	https > 59838 [FIN, ACK] Seq=91303 Ack=18373 Win=22619 Len=0

Figure 43. First 30 packets Captured on Host A During Test 2

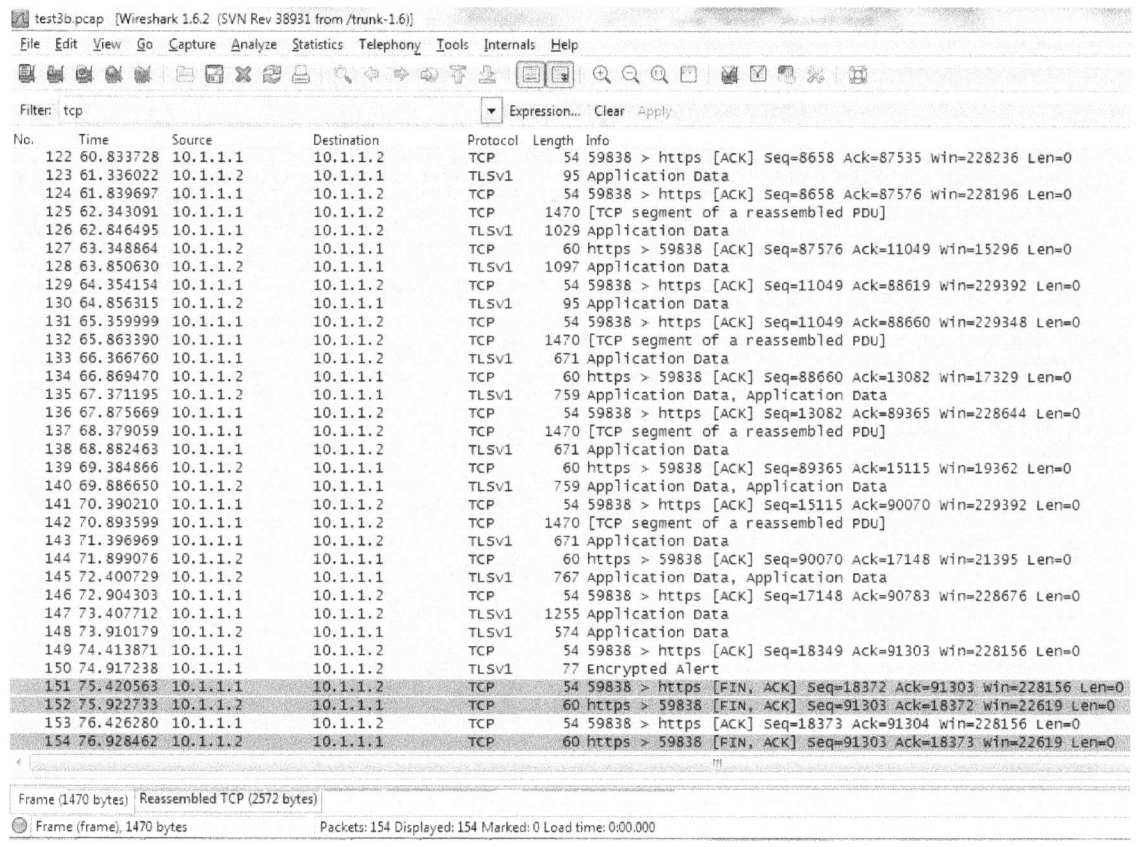

Figure 44. First 30 packets Captured on Host B During Test 2

Figure 46 shows a packet from the targeted TCP flow in the source network trace file that contains TCP data. Figures 47 and 48 show the same packet captured on Host A and Host B respectively, during the test. The IP and MAC addresses were changed during the test to match the addressing scheme of the test. The SEQ, ACK, TCP flags, window size, TCP segment data, and encrypted application data all match. This shows that the packets can be re-created correctly when used for testing.

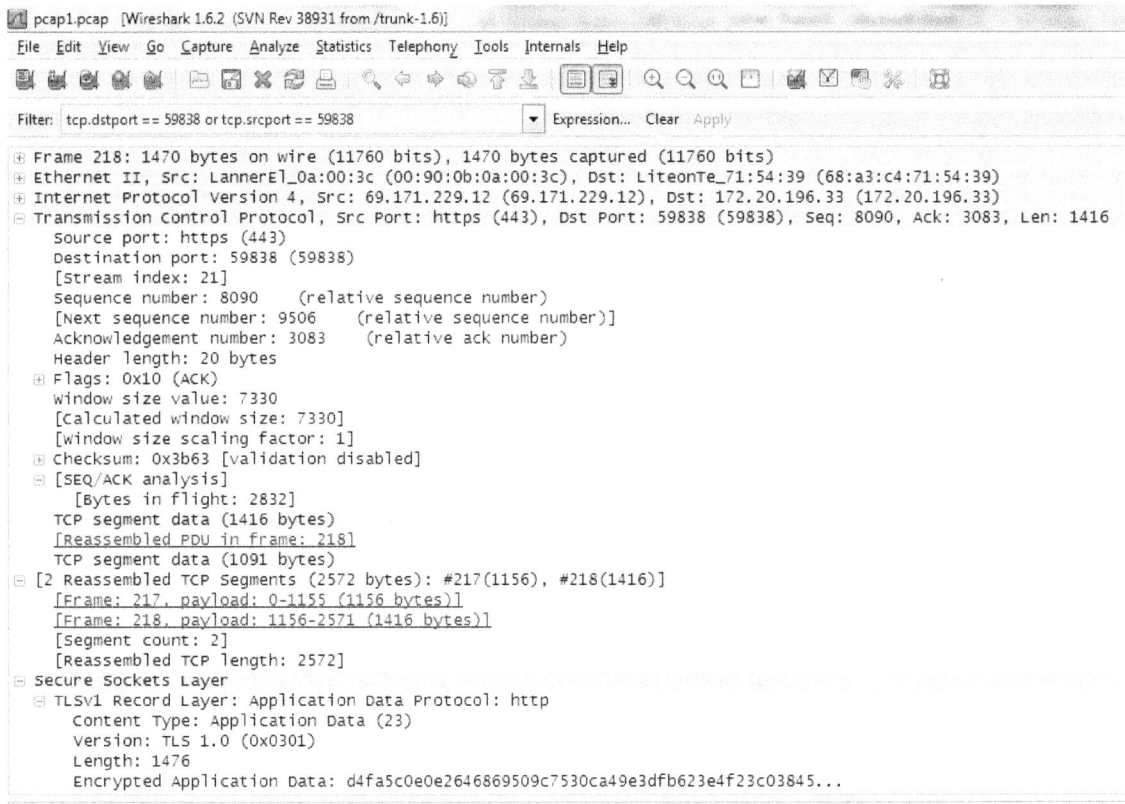

Figure 45.   Test 3 Packet With Data From Source Network
Trace File

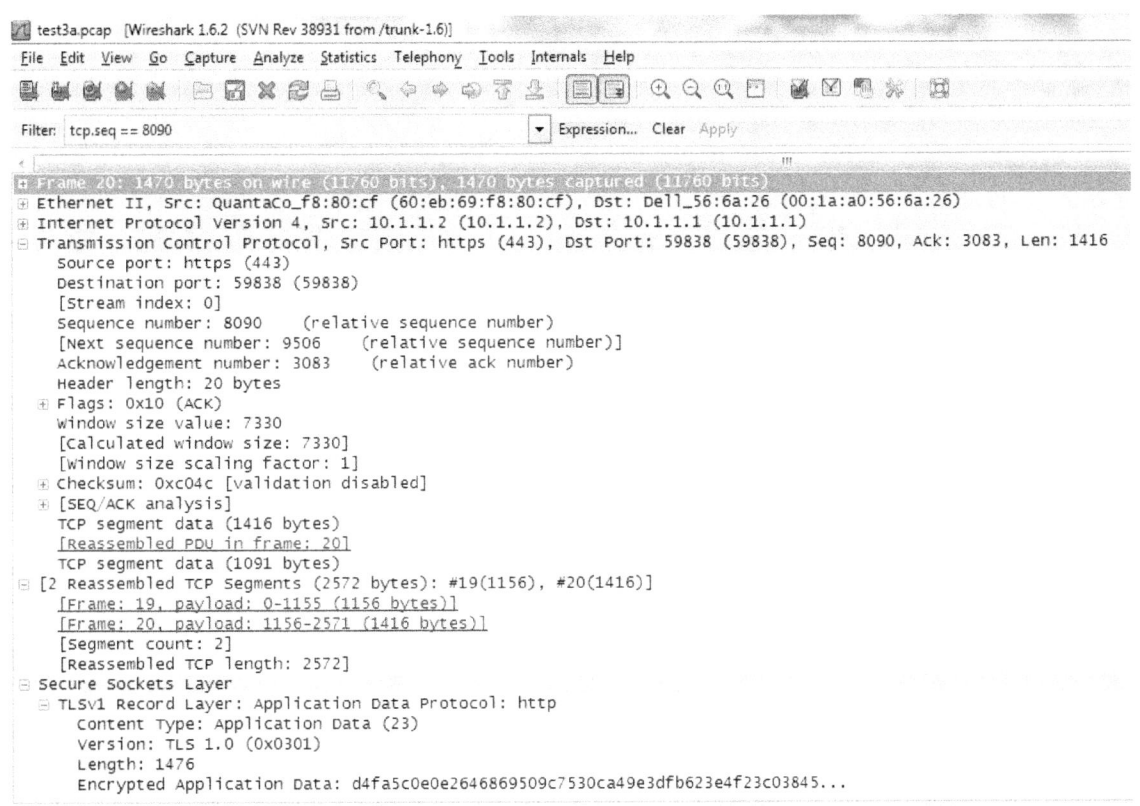

Figure 46.   Test 3 Packet With Data From Host A

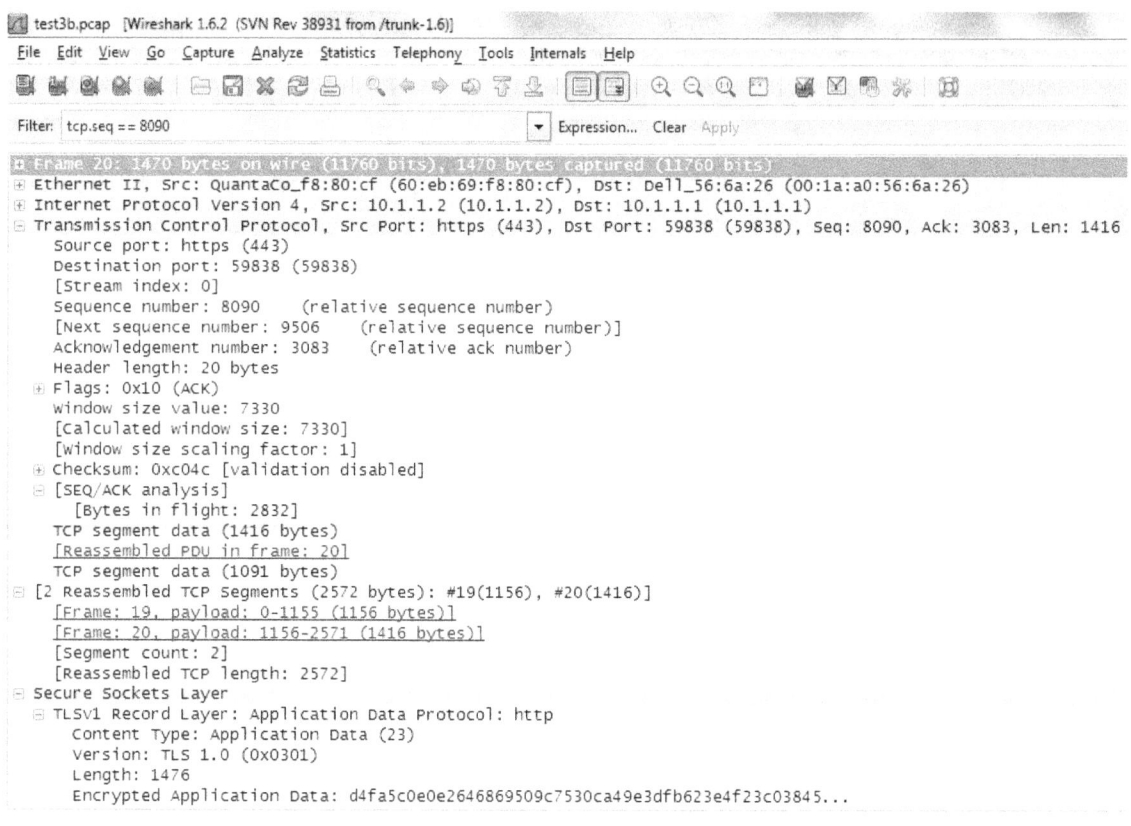

```
test3b.pcap  [Wireshark 1.6.2 (SVN Rev 38931 from /trunk-1.6)]
File  Edit  View  Go  Capture  Analyze  Statistics  Telephony  Tools  Internals  Help

Filter: tcp.seq == 8090                               ▼  Expression...  Clear  Apply

⊞ Frame 20: 1470 bytes on wire (11760 bits), 1470 bytes captured (11760 bits)
⊞ Ethernet II, Src: QuantaCo_f8:80:cf (60:eb:69:f8:80:cf), Dst: Dell_56:6a:26 (00:1a:a0:56:6a:26)
⊞ Internet Protocol Version 4, Src: 10.1.1.2 (10.1.1.2), Dst: 10.1.1.1 (10.1.1.1)
⊟ Transmission Control Protocol, Src Port: https (443), Dst Port: 59838 (59838), Seq: 8090, Ack: 3083, Len: 1416
     Source port: https (443)
     Destination port: 59838 (59838)
     [Stream index: 0]
     Sequence number: 8090    (relative sequence number)
     [Next sequence number: 9506    (relative sequence number)]
     Acknowledgement number: 3083    (relative ack number)
     Header length: 20 bytes
   ⊞ Flags: 0x10 (ACK)
     window size value: 7330
     [Calculated window size: 7330]
     [Window size scaling factor: 1]
   ⊞ Checksum: 0xc04c [validation disabled]
   ⊟ [SEQ/ACK analysis]
        [Bytes in flight: 2832]
     TCP segment data (1416 bytes)
     [Reassembled PDU in frame: 20]
     TCP segment data (1091 bytes)
⊟ [2 Reassembled TCP Segments (2572 bytes): #19(1156), #20(1416)]
     [Frame: 19, payload: 0-1155 (1156 bytes)]
     [Frame: 20, payload: 1156-2571 (1416 bytes)]
     [Segment count: 2]
     [Reassembled TCP length: 2572]
⊟ Secure Sockets Layer
   ⊟ TLSv1 Record Layer: Application Data Protocol: http
        Content Type: Application Data (23)
        Version: TLS 1.0 (0x0301)
        Length: 1476
        Encrypted Application Data: d4fa5c0e0e2646869509c7530ca49e3dfb623e4f23c03845...
```

Figure 47.   Test 3 Packet With Data From Host B

E.   TEST 4 – LARGE FILE DOWNLOAD FROM WEB

Test 4 was designed to further stress test
tcpFlowPrepper.py and *orderedReplay.py*. The source network
trace file was captured while downloading a large from the
Internet with several other web browser windows open. The
file download was approximately 208MB and would compromise
most of the 220MB source network trace file. This test was
designed to stress the database and the ability re-create
such a large flow. However, it overwhelmed the system
resources of the computer used to run *tcpFlowPrepper.py* and
caused the program to hang. Wireshark was used to truncate
the source network trace file to 110MB and
tcpFlowPrepper.py was run again. This was still too much
for *tcpFlowPrepper.py* to handle and the source network

trace file truncated to 48.5MB. This time *tcpFlowPrepper.py* was able to complete. Table 2 show that the entire source network trace file was 48.5MB and the targeted TCP flow made up 48.3MB of that. This shows that even though several other web browser windows were open, the targeted TCP flow dominates the source network file.

Table 4 shows how long it took to run each program for test 4 and highlights the need to optimize the code for *tcpFlowPrepper.py* and *orderedReplay.py*. It took *tcpFlowPrepper.py* 16 hours to complete with an empty database and one hour and 25 minutes to complete with a full database. It took *orderedReplay.py* 15 hours and seven minutes to complete sending all the packets. It took *flowStats.py* 3.2 seconds to complete gathering all the statistical data on the source network trace file and the targeted TCP flow.

Figure 48 shows the first 35 packets from the source network trace file and can be compared to Figure 49 and Figure 50, captured on each host during the test, to see that the captures match the source network trace in relation to the relative ordering of packets. Figure 51 shows the last 35 packets from the network source file and can be compared to Figure 52 and Figure 53, captured on each host during the test, to see that the captures match the source network trace in relation to the relative ordering of packets at the end of the test. This demonstrates that that *orderedReplay.py* was able to maintain the relative order of packets while replaying the targeted TCP flow. It would not be prudent to try and show a screen capture of the source network trace and each of the captures conducted during the test for over 50,000 packets so each of the three files were printed to a text file and compared manually with no discrepancies noted. This is a very time consuming process and underscores the need for an automated way to compare the network trace files.

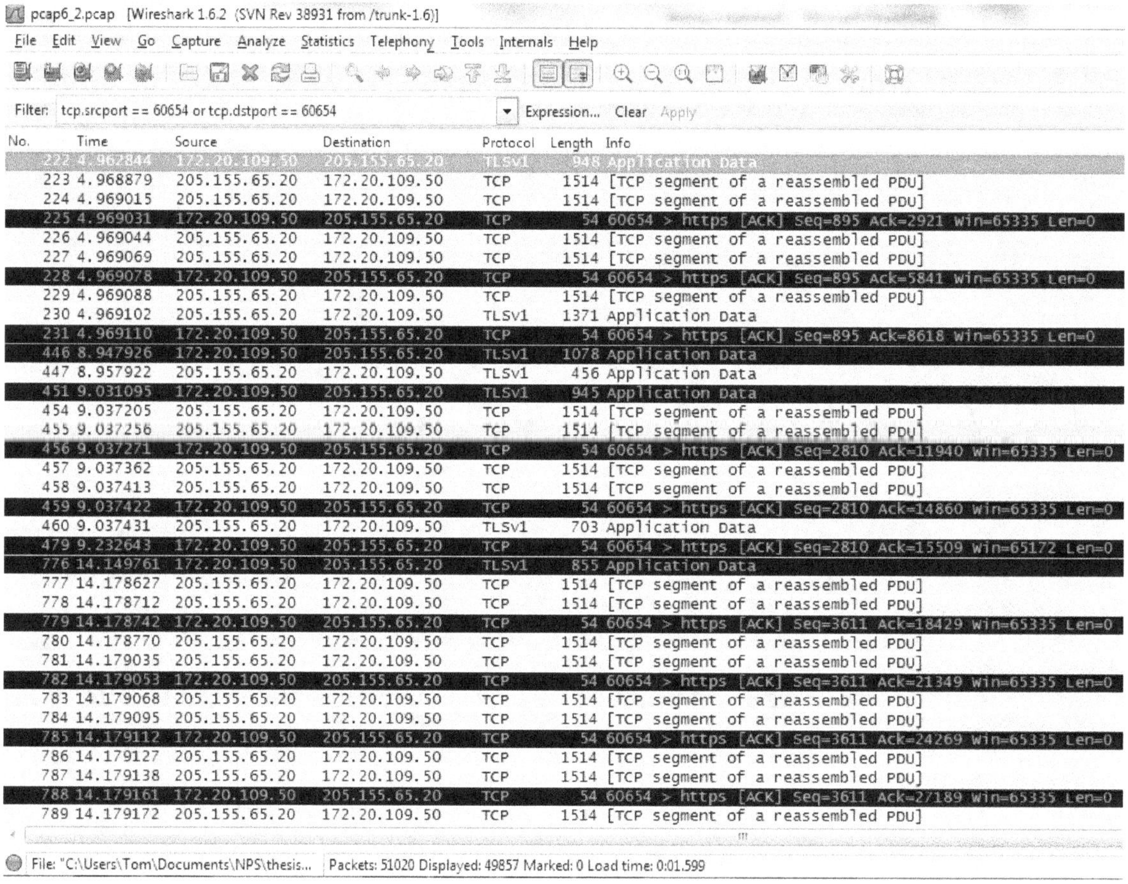

Figure 48.   First 35 Packets from Source Network
Trace File

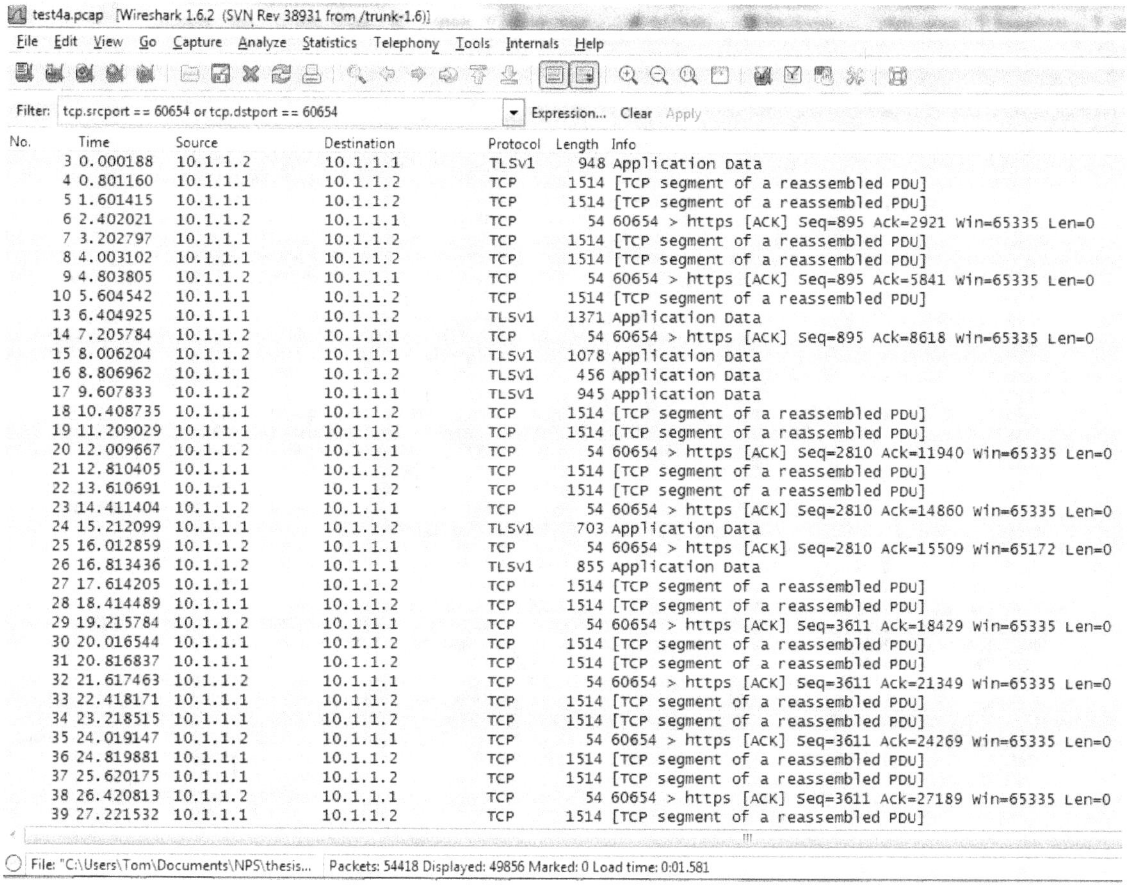

Figure 49.   First 35 Packets Captured at Host A
During Test

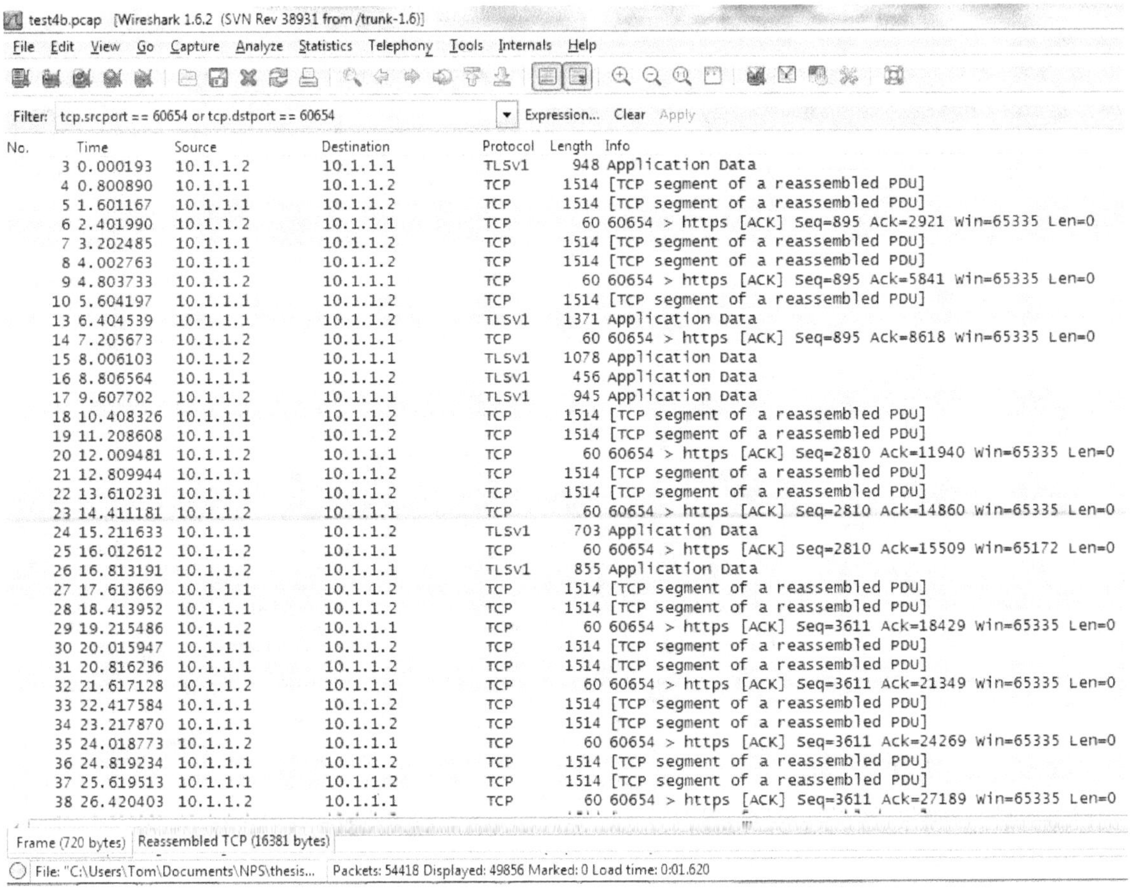

No.	Time	Source	Destination	Protocol	Length	Info
3	0.000193	10.1.1.2	10.1.1.1	TLSv1	948	Application Data
4	0.800890	10.1.1.1	10.1.1.2	TCP	1514	[TCP segment of a reassembled PDU]
5	1.601167	10.1.1.1	10.1.1.2	TCP	1514	[TCP segment of a reassembled PDU]
6	2.401990	10.1.1.2	10.1.1.1	TCP	60	60654 > https [ACK] Seq=895 Ack=2921 Win=65335 Len=0
7	3.202485	10.1.1.1	10.1.1.2	TCP	1514	[TCP segment of a reassembled PDU]
8	4.002763	10.1.1.1	10.1.1.2	TCP	1514	[TCP segment of a reassembled PDU]
9	4.803733	10.1.1.2	10.1.1.1	TCP	60	60654 > https [ACK] Seq=895 Ack=5841 Win=65335 Len=0
10	5.604197	10.1.1.1	10.1.1.2	TCP	1514	[TCP segment of a reassembled PDU]
13	6.404539	10.1.1.1	10.1.1.2	TLSv1	1371	Application Data
14	7.205673	10.1.1.2	10.1.1.1	TCP	60	60654 > https [ACK] Seq=895 Ack=8618 Win=65335 Len=0
15	8.006103	10.1.1.2	10.1.1.1	TLSv1	1078	Application Data
16	8.806564	10.1.1.1	10.1.1.2	TLSv1	456	Application Data
17	9.607702	10.1.1.2	10.1.1.1	TLSv1	945	Application Data
18	10.408326	10.1.1.1	10.1.1.2	TCP	1514	[TCP segment of a reassembled PDU]
19	11.208608	10.1.1.1	10.1.1.2	TCP	1514	[TCP segment of a reassembled PDU]
20	12.009481	10.1.1.2	10.1.1.1	TCP	60	60654 > https [ACK] Seq=2810 Ack=11940 Win=65335 Len=0
21	12.809944	10.1.1.1	10.1.1.2	TCP	1514	[TCP segment of a reassembled PDU]
22	13.610231	10.1.1.1	10.1.1.2	TCP	1514	[TCP segment of a reassembled PDU]
23	14.411181	10.1.1.2	10.1.1.1	TCP	60	60654 > https [ACK] Seq=2810 Ack=14860 Win=65335 Len=0
24	15.211633	10.1.1.1	10.1.1.2	TLSv1	703	Application Data
25	16.012612	10.1.1.2	10.1.1.1	TCP	60	60654 > https [ACK] Seq=2810 Ack=15509 Win=65172 Len=0
26	16.813191	10.1.1.2	10.1.1.1	TLSv1	855	Application Data
27	17.613669	10.1.1.1	10.1.1.2	TCP	1514	[TCP segment of a reassembled PDU]
28	18.413952	10.1.1.1	10.1.1.2	TCP	1514	[TCP segment of a reassembled PDU]
29	19.215486	10.1.1.2	10.1.1.1	TCP	60	60654 > https [ACK] Seq=3611 Ack=18429 Win=65335 Len=0
30	20.015947	10.1.1.1	10.1.1.2	TCP	1514	[TCP segment of a reassembled PDU]
31	20.816236	10.1.1.1	10.1.1.2	TCP	1514	[TCP segment of a reassembled PDU]
32	21.617128	10.1.1.2	10.1.1.1	TCP	60	60654 > https [ACK] Seq=3611 Ack=21349 Win=65335 Len=0
33	22.417584	10.1.1.1	10.1.1.2	TCP	1514	[TCP segment of a reassembled PDU]
34	23.217870	10.1.1.1	10.1.1.2	TCP	1514	[TCP segment of a reassembled PDU]
35	24.018773	10.1.1.2	10.1.1.1	TCP	60	60654 > https [ACK] Seq=3611 Ack=24269 Win=65335 Len=0
36	24.819234	10.1.1.1	10.1.1.2	TCP	1514	[TCP segment of a reassembled PDU]
37	25.619513	10.1.1.1	10.1.1.2	TCP	1514	[TCP segment of a reassembled PDU]
38	26.420403	10.1.1.2	10.1.1.1	TCP	60	60654 > https [ACK] Seq=3611 Ack=27189 Win=65335 Len=0

Figure 50. First 35 Packets Captured at Host B During Test

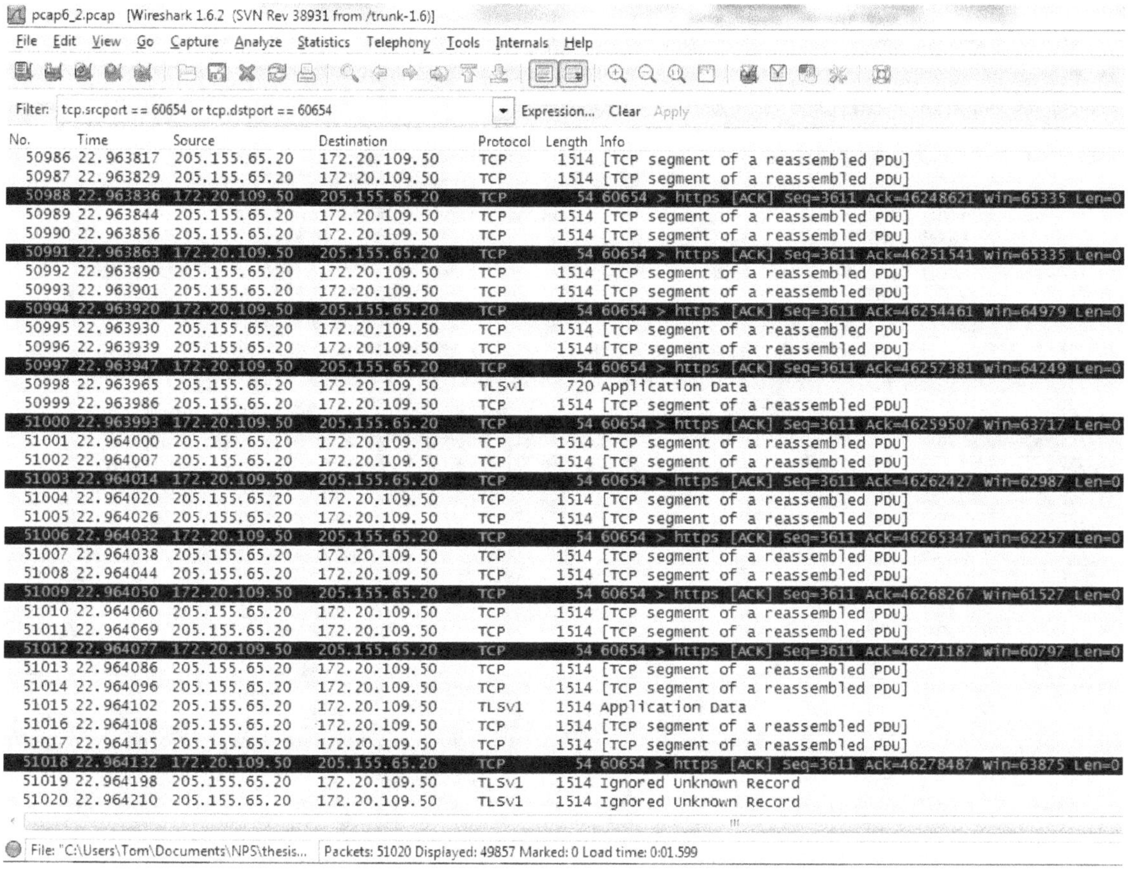

No.	Time	Source	Destination	Protocol	Length	Info
50986	22.963817	205.155.65.20	172.20.109.50	TCP	1514	[TCP segment of a reassembled PDU]
50987	22.963829	205.155.65.20	172.20.109.50	TCP	1514	[TCP segment of a reassembled PDU]
50988	22.963836	172.20.109.50	205.155.65.20	TCP	54	60654 > https [ACK] Seq=3611 Ack=46248621 Win=65335 Len=0
50989	22.963844	205.155.65.20	172.20.109.50	TCP	1514	[TCP segment of a reassembled PDU]
50990	22.963856	205.155.65.20	172.20.109.50	TCP	1514	[TCP segment of a reassembled PDU]
50991	22.963863	172.20.109.50	205.155.65.20	TCP	54	60654 > https [ACK] Seq=3611 Ack=46251541 Win=65335 Len=0
50992	22.963890	205.155.65.20	172.20.109.50	TCP	1514	[TCP segment of a reassembled PDU]
50993	22.963901	205.155.65.20	172.20.109.50	TCP	1514	[TCP segment of a reassembled PDU]
50994	22.963920	172.20.109.50	205.155.65.20	TCP	54	60654 > https [ACK] Seq=3611 Ack=46254461 Win=64979 Len=0
50995	22.963930	205.155.65.20	172.20.109.50	TCP	1514	[TCP segment of a reassembled PDU]
50996	22.963939	205.155.65.20	172.20.109.50	TCP	1514	[TCP segment of a reassembled PDU]
50997	22.963947	172.20.109.50	205.155.65.20	TCP	54	60654 > https [ACK] Seq=3611 Ack=46257381 Win=64249 Len=0
50998	22.963965	205.155.65.20	172.20.109.50	TLSv1	720	Application Data
50999	22.963986	205.155.65.20	172.20.109.50	TCP	1514	[TCP segment of a reassembled PDU]
51000	22.963993	172.20.109.50	205.155.65.20	TCP	54	60654 > https [ACK] Seq=3611 Ack=46259507 Win=63717 Len=0
51001	22.964000	205.155.65.20	172.20.109.50	TCP	1514	[TCP segment of a reassembled PDU]
51002	22.964007	205.155.65.20	172.20.109.50	TCP	1514	[TCP segment of a reassembled PDU]
51003	22.964014	172.20.109.50	205.155.65.20	TCP	54	60654 > https [ACK] Seq=3611 Ack=46262427 Win=62987 Len=0
51004	22.964020	205.155.65.20	172.20.109.50	TCP	1514	[TCP segment of a reassembled PDU]
51005	22.964026	205.155.65.20	172.20.109.50	TCP	1514	[TCP segment of a reassembled PDU]
51006	22.964032	172.20.109.50	205.155.65.20	TCP	54	60654 > https [ACK] Seq=3611 Ack=46265347 Win=62257 Len=0
51007	22.964038	205.155.65.20	172.20.109.50	TCP	1514	[TCP segment of a reassembled PDU]
51008	22.964044	205.155.65.20	172.20.109.50	TCP	1514	[TCP segment of a reassembled PDU]
51009	22.964050	172.20.109.50	205.155.65.20	TCP	54	60654 > https [ACK] Seq=3611 Ack=46268267 Win=61527 Len=0
51010	22.964060	205.155.65.20	172.20.109.50	TCP	1514	[TCP segment of a reassembled PDU]
51011	22.964069	205.155.65.20	172.20.109.50	TCP	1514	[TCP segment of a reassembled PDU]
51012	22.964077	172.20.109.50	205.155.65.20	TCP	54	60654 > https [ACK] Seq=3611 Ack=46271187 Win=60797 Len=0
51013	22.964086	205.155.65.20	172.20.109.50	TCP	1514	[TCP segment of a reassembled PDU]
51014	22.964096	205.155.65.20	172.20.109.50	TCP	1514	[TCP segment of a reassembled PDU]
51015	22.964102	205.155.65.20	172.20.109.50	TLSv1	1514	Application Data
51016	22.964108	205.155.65.20	172.20.109.50	TCP	1514	[TCP segment of a reassembled PDU]
51017	22.964115	205.155.65.20	172.20.109.50	TCP	1514	[TCP segment of a reassembled PDU]
51018	22.964132	172.20.109.50	205.155.65.20	TCP	54	60654 > https [ACK] Seq=3611 Ack=46278487 Win=63875 Len=0
51019	22.964198	205.155.65.20	172.20.109.50	TLSv1	1514	Ignored Unknown Record
51020	22.964210	205.155.65.20	172.20.109.50	TLSv1	1514	Ignored Unknown Record

Figure 51. Last 35 Packets from Source Network Trace File

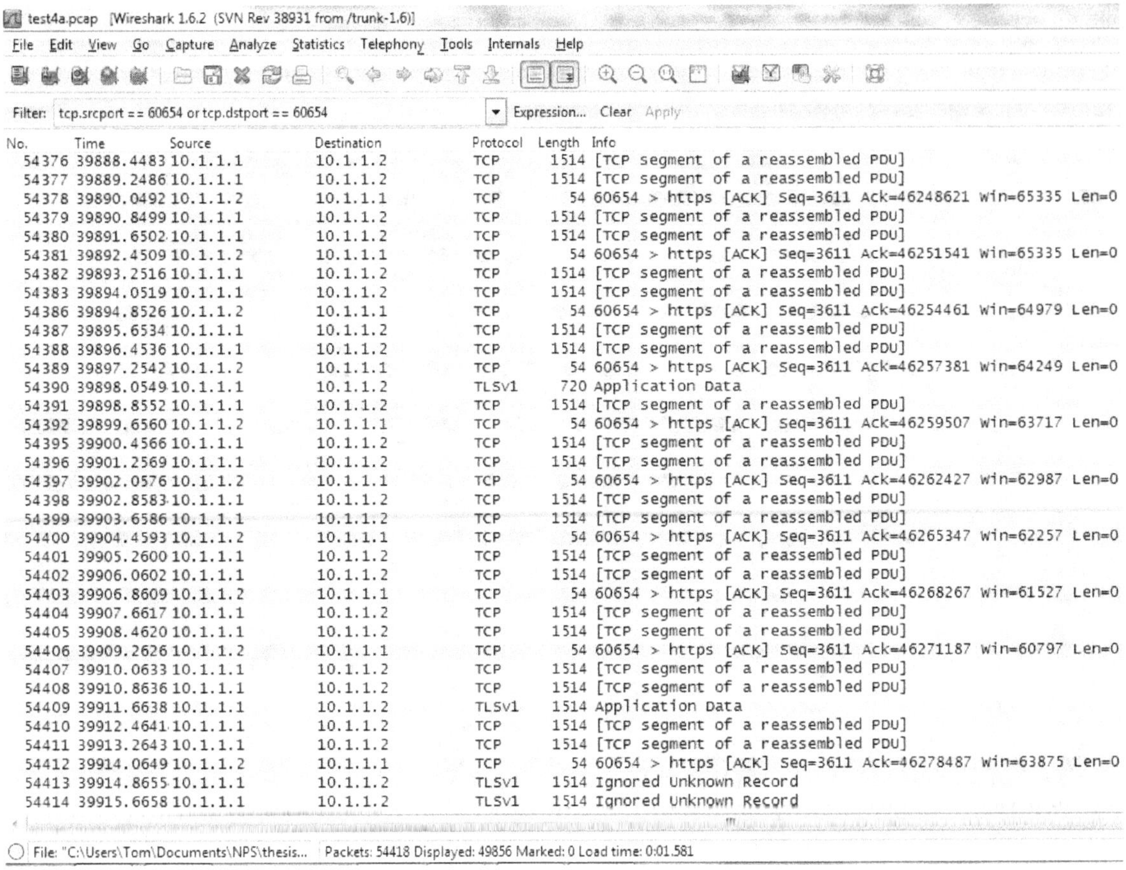

No.	Time	Source	Destination	Protocol	Length	Info
54376	39888.4483	10.1.1.1	10.1.1.2	TCP	1514	[TCP segment of a reassembled PDU]
54377	39889.2486	10.1.1.1	10.1.1.2	TCP	1514	[TCP segment of a reassembled PDU]
54378	39890.0492	10.1.1.2	10.1.1.1	TCP	54	60654 > https [ACK] Seq=3611 Ack=46248621 Win=65335 Len=0
54379	39890.8499	10.1.1.1	10.1.1.2	TCP	1514	[TCP segment of a reassembled PDU]
54380	39891.6502	10.1.1.1	10.1.1.2	TCP	1514	[TCP segment of a reassembled PDU]
54381	39892.4509	10.1.1.2	10.1.1.1	TCP	54	60654 > https [ACK] Seq=3611 Ack=46251541 Win=65335 Len=0
54382	39893.2516	10.1.1.1	10.1.1.2	TCP	1514	[TCP segment of a reassembled PDU]
54383	39894.0519	10.1.1.1	10.1.1.2	TCP	1514	[TCP segment of a reassembled PDU]
54386	39894.8526	10.1.1.2	10.1.1.1	TCP	54	60654 > https [ACK] Seq=3611 Ack=46254461 Win=64979 Len=0
54387	39895.6534	10.1.1.1	10.1.1.2	TCP	1514	[TCP segment of a reassembled PDU]
54388	39896.4536	10.1.1.1	10.1.1.2	TCP	1514	[TCP segment of a reassembled PDU]
54389	39897.2542	10.1.1.2	10.1.1.1	TCP	54	60654 > https [ACK] Seq=3611 Ack=46257381 Win=64249 Len=0
54390	39898.0549	10.1.1.1	10.1.1.2	TLSv1	720	Application Data
54391	39898.8552	10.1.1.1	10.1.1.2	TCP	1514	[TCP segment of a reassembled PDU]
54392	39899.6560	10.1.1.2	10.1.1.1	TCP	54	60654 > https [ACK] Seq=3611 Ack=46259507 Win=63717 Len=0
54395	39900.4566	10.1.1.1	10.1.1.2	TCP	1514	[TCP segment of a reassembled PDU]
54396	39901.2569	10.1.1.1	10.1.1.2	TCP	1514	[TCP segment of a reassembled PDU]
54397	39902.0576	10.1.1.2	10.1.1.1	TCP	54	60654 > https [ACK] Seq=3611 Ack=46262427 Win=62987 Len=0
54398	39902.8583	10.1.1.1	10.1.1.2	TCP	1514	[TCP segment of a reassembled PDU]
54399	39903.6586	10.1.1.1	10.1.1.2	TCP	1514	[TCP segment of a reassembled PDU]
54400	39904.4593	10.1.1.2	10.1.1.1	TCP	54	60654 > https [ACK] Seq=3611 Ack=46265347 Win=62257 Len=0
54401	39905.2600	10.1.1.1	10.1.1.2	TCP	1514	[TCP segment of a reassembled PDU]
54402	39906.0602	10.1.1.1	10.1.1.2	TCP	1514	[TCP segment of a reassembled PDU]
54403	39906.8609	10.1.1.2	10.1.1.1	TCP	54	60654 > https [ACK] Seq=3611 Ack=46268267 Win=61527 Len=0
54404	39907.6617	10.1.1.1	10.1.1.2	TCP	1514	[TCP segment of a reassembled PDU]
54405	39908.4620	10.1.1.1	10.1.1.2	TCP	1514	[TCP segment of a reassembled PDU]
54406	39909.2626	10.1.1.2	10.1.1.1	TCP	54	60654 > https [ACK] Seq=3611 Ack=46271187 Win=60797 Len=0
54407	39910.0633	10.1.1.1	10.1.1.2	TCP	1514	[TCP segment of a reassembled PDU]
54408	39910.8636	10.1.1.1	10.1.1.2	TCP	1514	[TCP segment of a reassembled PDU]
54409	39911.6638	10.1.1.1	10.1.1.2	TLSv1	1514	Application Data
54410	39912.4641	10.1.1.1	10.1.1.2	TCP	1514	[TCP segment of a reassembled PDU]
54411	39913.2643	10.1.1.1	10.1.1.2	TCP	1514	[TCP segment of a reassembled PDU]
54412	39914.0649	10.1.1.2	10.1.1.1	TCP	54	60654 > https [ACK] Seq=3611 Ack=46278487 Win=63875 Len=0
54413	39914.8655	10.1.1.1	10.1.1.2	TLSv1	1514	Ignored Unknown Record
54414	39915.6658	10.1.1.1	10.1.1.2	TLSv1	1514	Ignored Unknown Record

Figure 52. Last 35 Packets Captured from Host A

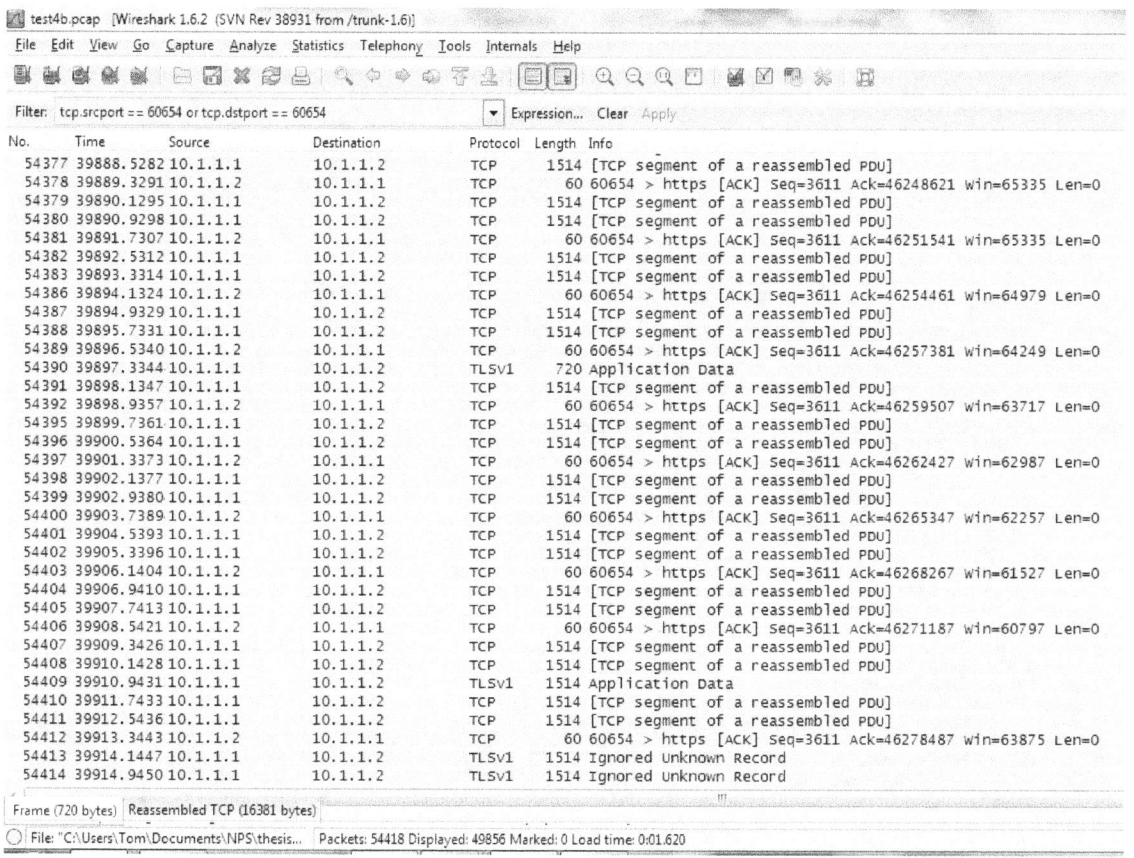

Figure 53.   Last 35 Packets Captured on Host B

Figure 54 shows a TCP data packet with application data from the source network trace. Figure 55 and Figure 56 show the same TCP packet from the targeted TCP flow captured on each host during the test. The source and destination MAC and IP addresses were changed to match the testing scheme, as a result they do not match between the captures during testing and the targeted TCP flow in the source network trace file. The following packet traits do match between the packet from the source network trace and from the packet captured during testing: source port number, destination port number, the sequence number, the acknowledgment number, the packet size, the header length, the TCP options, the TCP flags, and the TCP data. This

103

demonstrates that *orderedReplay.py* successfully changed the address scheme to match the testing scheme but kept all other packet traits from the source network trace in the re-created TCP flow.

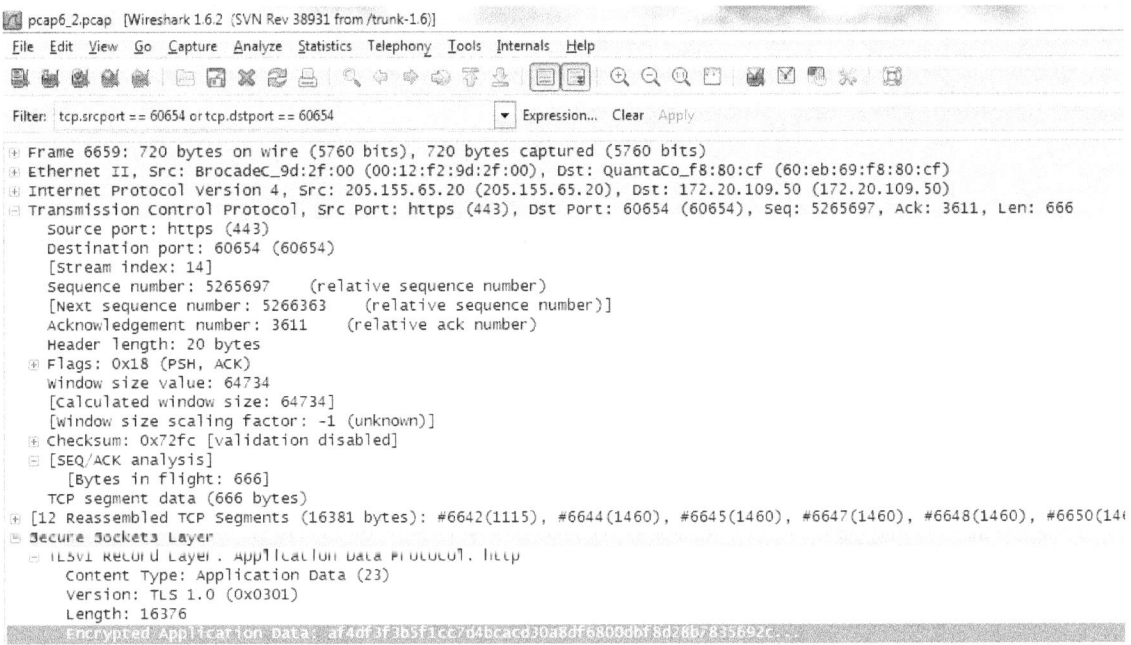

Figure 54.   TCP Packet with Application Data from Source
Network Trace File

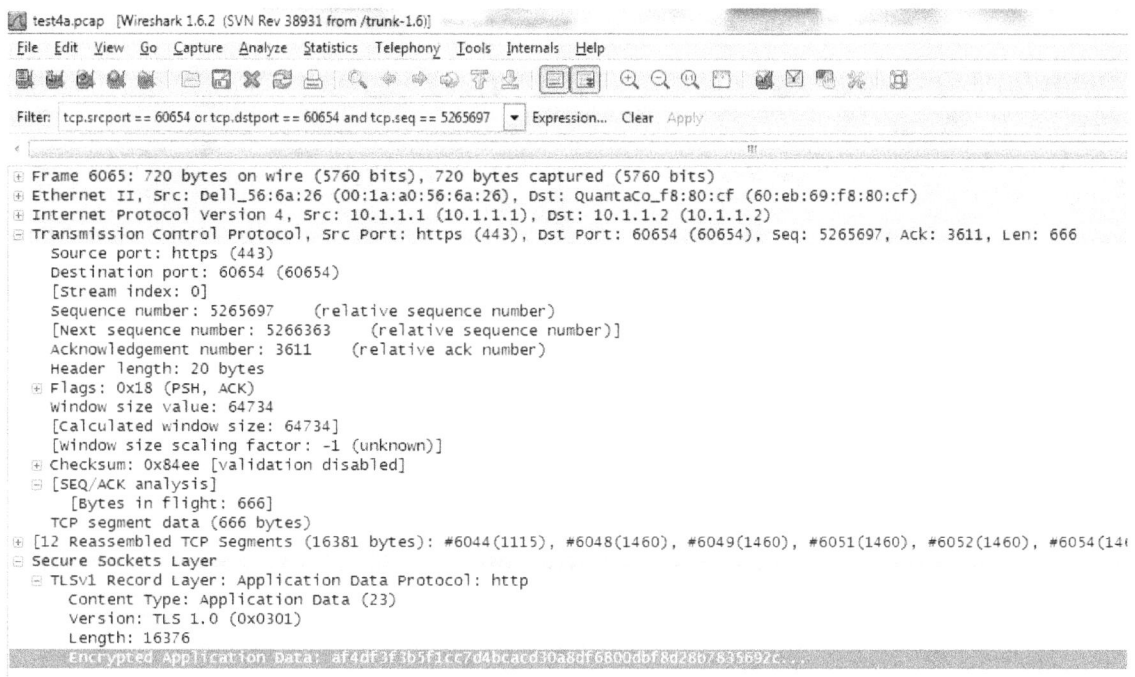

Figure 55. TCP packet with Application Data Captured
on Host A

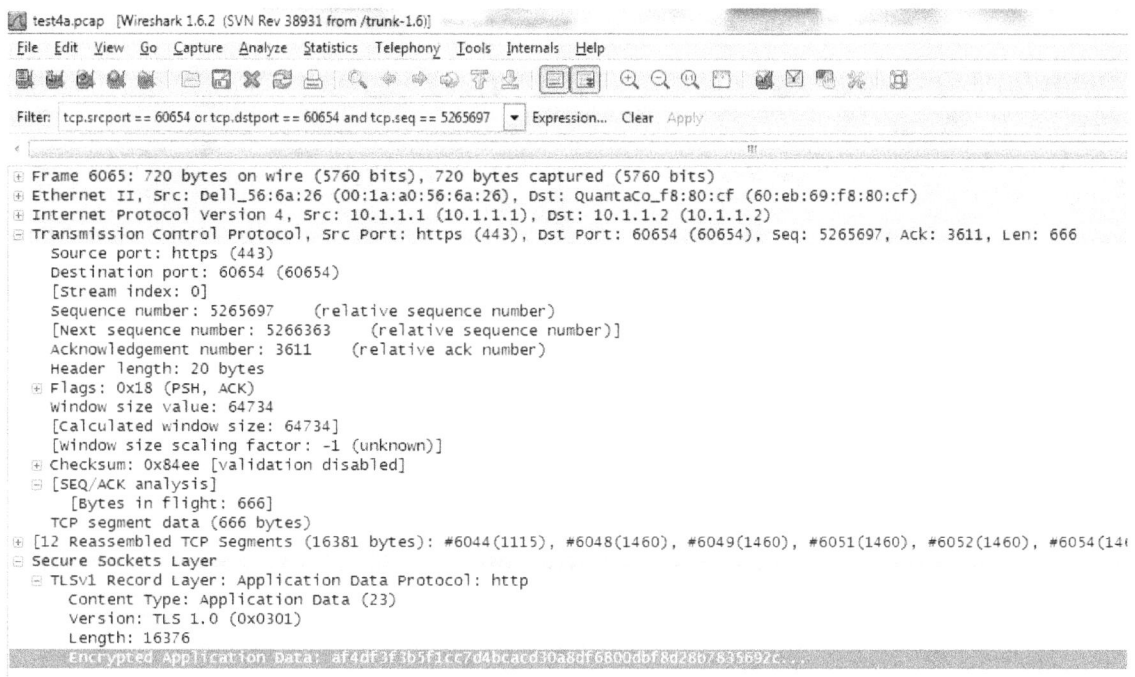

Figure 56. TCP Packet with Application Data Captured
on Host B

In conclusion, this chapter demonstrates that the tool created allowed two hosts to replay a specific TCP flow without retransmissions from a source network trace file in the same order as the original TCP flow without altering the TCP payload. The program also allowed the headers of the packets being sent to be modified to the addressing scheme of the testing setup. The main technical issues discovered were: *tcpFlowPrepper.py* would hang for source network trace file that were larger than 50MB, there is a possibility of deadlock when using *orderedReplay.py* if the hosts get out of synchronization, *orderedReplay.py* has a significant delay between sending packets in order to maintain synchronization, and manual inspection to compare the source network trace files and the files captured during testing becomes unmanageable as the target TCP flow becomes very large. These issues will be addressed in the conclusion and future works section.

V.   CONCLUSIONS, RECOMMENDATIONS, AND FUTURE WORK

A.   CONCLUSIONS

This thesis laid the groundwork and establishes a framework for the development of a testing capability for measuring the efficiency of different WAN optimization products with respect to an excerpt of real-world production TCP flows. It demonstrated the ability of the tool to successfully read a network trace file, populate the database with IP packet traits, and mark retransmitted packets. It also demonstrated the repeatability of the test by being able to re-create different TCP flows from the same network trace file.

There were several issues discovered during testing that need to be addressed in order to move forward and meet the requirements stated by MCTSSA. First, the *tcpFlowPrepper.py* hangs and does not complete when attempting to mark retransmissions if the database gets larger than 50,000 records. This limits testing to the use of orderedReplay.py on files up to 50MB. There was significant delay in completing the re-creation of the file transfer in order to ensure the packets were transmitted with the proper sequencing between hosts. If the packets do get out of sequence then *orderedReplay.py* will transition into a deadlock state where each host is waiting for the other host to transmit a packet(s).

The program *flowStats.py* provides good statistical data on a network trace file and a particular TCP flow within that trace file. However, at this time there is not an automated way to compare packets within a TCP flow of the source network trace file and the network trace files captured during testing. All comparisons must be performed manually, which is time consuming and may be error prone.

B.   RECOMMENDATIONS

Recommend that both hosts used in testing are the same platform (i.e., both hosts are Dell Optiplex 745 Desktop), have the same processing power and RAM, and use the same Operating System and version. This would eliminate the disparity causing the packet sizes to be different between Host A and Host B during testing caused by one host operating system stripping the Ethernet trailer and the other host retaining it. It would also ensure that both hosts have the same performance to ensure that one host is not a bottle-neck for testing.

It is recommended that the source code programs, *tcpFLowPrepper.py* and *orderedReplay.py*, be reviewed for optimization to prevent overwhelming the system resources. It is also recommended that the MySQL queries be reviewed for efficiency with respect to database access to reduce processing time and to prevent *tcpFlowPrepper.py* from hanging when working with large TCP flows. Further inspection of SQL queries in *tcpFlowPrepper.py* revealed that large amounts of data were being retrieved using the cursor.fetchall() method, which returns all of the results at once. In future versions, it may be more efficient to use a loop and fetch one result at a time. However, this remains to be verified. In order to significantly improve the performance of *tcpFlowPrepper.py*, it might also be helpful to look into non-relational databases to see if they can improve the performance of adding packet data and retrieving queries

One of the goals of the tool produced by this thesis effort was to maintain the same inter-packet timing when a TCP flow is replayed as it had in the TCP flow within the original source network trace file. The pause required before sending a packet is a significant part of this issue and is also a significant reason for long run-time on tests for *orderedReplay.py*. It is recommended the deadlock issue be addressed so that deadlock is avoided.

The program *flowStats.py* should be expanded to automate the comparison of packets between the target flow(s) in the source network trace and the network trace files captured during testing with orderedReplay.py. This will significantly reduce the workload and increase the accuracy in verifying that TCP packets captured during testing match TCP packets from the source network trace file.

C.   FUTURE WORK

The software tool created in this thesis performs ordered replay and lays the ground work to create a tool that performs stateful TCP replay for performance evaluation of WAN optimization products. In order to do this, more research needs to be performed on how WAN optimization products perform optimization to determine whether they optimize a single flow or multiple flows over the troubled link. More research also needs to be performed in the area of stateful TCP replay to determine if it is possible and for determining the best way to implement it.

The testing network architecture needs to be implemented (See Figure 58). The testing scheme will require two servers capable of running WAN optimization virtual machines. WAN optimization virtual machine licenses will need to be obtained from at least two vendors. A WAN emulator will need to be setup to recreate network conditions such as average round trip time, bit error rate, and transmission rate. Finally, the testing scheme will require real-world production network trace files for testing.

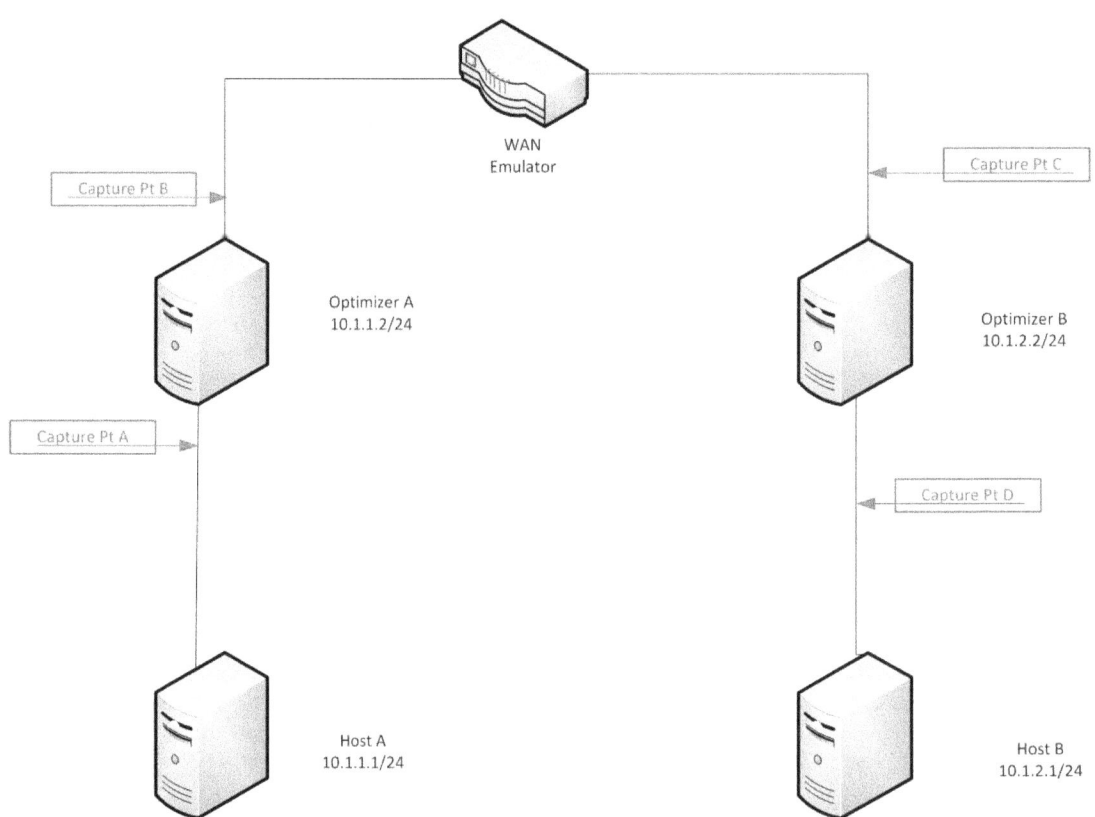

Figure 57.   WAN Optimization Testing Architecture

APPENDIX

A.   SOURCE CODE FOR TCPFLOWPREPPER.PY

```
##   Author:  Tom LeVier
##   Date:  23 March 2013
##   File Name: StcpFlowPrepper.py
##   Masters Thesis
##   "A Tool for Stateful TCP Replay"
##   Thesis advisors:
##      Professor Xie
##      Professor Gibson
##
##   Purpose:  The purpose of this program is to
##   read in a .pcap file.  Packets are first
##   entered into an MySQL DB and retransmissions
##   are marked.  Then a target TCP flow is
##   identified using an input text file.  Packets
##   that match the flow are written to a new .pcap
##   based on sending host.  Retransmissions are
##   not written to the new file. The order of
##   sending is written to a file by host name.
##
##   Uses Python 2.7 on ubuntu 11.10


import sys
import dpkt
import pcap
import socket
import binascii
import struct
import MySQLdb as sql
import array
import hashlib
from decimal import Decimal


#Global Variables

##track number of tcp packets
counter = 0

#port number to extract flow from
target_portNo = 0

#default src/dst ip and mac addresses
#get changed via user input file
src1 = '172.20.196.34'
dst1 = '198.189.255.202'
src_mac1 = '68:a3:c4:71:54:00'
dst_mac1 = '00:aa:11:aa:11:00'
```

```python
##  This splits up an IP address and converts it from in to char
##  so it is readable
def ipv4addr(addr):
    addrcode = [chr(int(i)) for i in addr.split('.')]
    return "".join(addrcode)


def main():
    global counter, target_portNo, src_ip, dst_ip, src_mac, dst_mac
    global packetHash, con, cursor, tcp, optsDict, pidList

    #source pcap file to read in
    filename = "pcap6_2.pcap"

    notIPcount = 0 # track number of packets that are not IP

    #create empty list for seq numbers
    seqList = [-1]

    #store the packet Id #s that for packets being
    #written to a new file
    pidList = []

     # connect to the database
    con3 = sql.connect("localhost", "root", "Dipper01", "pcapParser")
    cursor3 = con3.cursor()
    #make backslashes ok for DB
    cursor3.execute('set global sql_mode="NO_BACKSLASH_ESCAPES"')
    #disconnect from DB
    cursor3.close()
    con3.close()

     # check to see if database is populated
    con4 = sql.connect("localhost", "root", "Dipper01", "pcapParser")
    cursor4 = con4.cursor()
    message = cursor4.execute('select * from rawPacketData')
    #disconnect from DB
    cursor4.close()
    con4.close()

    print "opening .pcap to read"

    #open capture file to read
    pcr = dpkt.pcap.Reader(open(filename))

    #initialize packetID
    packetID = 1

    # only read the pcap file if the database is empty
    if message == 0:
        print "putting packets in database"

        ##  read in packet data using timestamp and buffer
        for ts, buf in pcr:
            #read ethernet data from the buffer
```

```python
            eth_in = dpkt.ethernet.Ethernet(buf)

            #check mac addresses from input file
            rdst_mac = decode_mac(eth_in.dst)
            rsrc_mac = decode_mac(eth_in.src)


            # determine if packet has IP data
            if eth_in.type == 2048:


                #save IP data
                ip_in = eth_in.data
                ip_data = binascii.b2a_base64(bytes(ip_in.data))

                try:
                    #insert all IP packets into the DB
                    stamp = Decimal(ts) #maintain precision of ts
                    #insert IP data for packet into DB
                    callInsertIP(packetID,  stamp,  rsrc_mac,  rdst_mac,
ip_in.src, ip_in.dst, ip_in.p, ip_data)
                except Exception as e:
                    print "error"
                    print packetID
                    print ip_data
                    print e


                #if IP data has TCP
                if ip_in.p == dpkt.ip.IP_PROTO_TCP:
                    #encapsulate TCP in IP
                    tcp = ip_in.data

                    #get src and dst ports
                    sport = tcp.sport
                    dport = tcp.dport

                    #save tcp data to put in db
                    options    =    binascii.b2a_base64(bytes(tcp.opts))
#convert to ascii for DB
                    seqNo = tcp.seq
                    ackNo = tcp.ack
                    flags = tcp.flags
                    recvWin = tcp.win
                    isReTran = False
                    tcpData    =    binascii.b2a_base64(bytes(tcp.data))
#convert to ascii for DB

                    try:
                        #add tcp data to db
                        callInsertTCP(packetID, flags, isReTran, sport,
dport, seqNo, ackNo, recvWin, tcpData,  str(options))
                    except Exception as e:
                        print "error tcp"
                        print packetID
                        print tcpData
```

```
                         print e
                 counter += 1 #track number of tcp packets

             else:
                 #not ip packet
                 notIPcount = notIPcount + 1
                 print "not ipPacket", packetID, ",   ", eth_in.type
##        #increment packetID
             packetID = packetID + 1
        print "\nFound ", notIPcount, " that were not IP packets"
        print str(packetID - notIPcount) + " were IP packets."


    try:
        print "marking duplicates"
        seqNoList = [] #list of distinct sequence numbers
        pktNoList = [] #list of pkt id's for a repeated sequence number

        #get a list of seqNo for tcp packets
        seqNoList = callGetSeqNo()

            #check each sequence number to see if it has
            # more than 1 pktID and len(tcpData) > 1
        for item in seqNoList:
##          creates a list of pktIDs if a seqNo has more than 1 pktID
##          with tcpData length > 1
            pktNoList = callFindDupes(item)

            #check to see if there is something in the lsit
            if len(pktNoList) > 1:

                #sort the list and remove the first packet
                #so it doesn't get marked
                pktNoList.sort()
                pktNoList.pop(0)

                #mark each packet in list as retx
                for stuff in pktNoList:
                    callMarkDupes(stuff[:-1])

    except Exception as e:
        print e

    #try reading parameters in from a file and save as variables
    #use these to write src/dst mac/ip to new pcap
    #files to match lab set up mac/ip addresses
    tgtfile = open("flow.txt", "r")
    tgt_sip = tgtfile.readline()
    tgt_sip = tgt_sip[:-1]
    tgt_sport = tgtfile.readline()
    tgt_sport = tgt_sport[:-1]
    tgt_dip = tgtfile.readline()
    tgt_dip = tgt_dip[:-1]
    tgt_dport = tgtfile.readline()
    tgt_dport = tgt_dport[:-1]
```

```
print "writing new pcap based on sending IP"
print "tgt_srcIP =", tgt_sip
print "tgt_src_port =", tgt_sport
print "dst_ip =", tgt_dip
print "tgt_dst_port =", tgt_dport



#retrives a list of packets for matching TCP flow for hostA
flowListA = getTgtFlow(tgt_sip, tgt_sport, tgt_dip, tgt_dport)

#retrives a list of packets for matching TCP flow for hostB
flowListB = getTgtFlow(tgt_dip, tgt_dport, tgt_sip, tgt_sport)

print "writing hosta.pcap"
writeHost(flowListA, 'A') #write packets sent by hostA
print "writing hostB.pcap"
writeHost(flowListB, 'B') #write packets sent by hostB

    #write the order of the packets to a file
con7 = sql.connect("localhost", "root", "Dipper01", "pcapParser")
cursor7 = con7.cursor()

#query to find all packets that are not retransmissions
# in the tcp flow we are looking for
string1 = "select ip.srcIP from TCPPacketData tcp, rawPacketData ip
where ((tcp.srcPort = " + str(tgt_dport) + " and tcp.dstPort = " +
str(tgt_sport) + " and ip.dstIP = '" + tgt_sip + "' and ip.srcIP = '" +
tgt_dip + "') or (tcp.srcPort = " + str(tgt_sport) + " and tcp.dstPort
= " + str(tgt_dport) + " and ip.srcIP = '" + tgt_sip + "' and ip.dstIP
= '" + tgt_dip + "')) and tcp.isRetrans = 0 and ip.proto = 6 and ip.pid
= tcp.pktID"
    msg = cursor7.execute(string1)
    ipString = str(cursor7.fetchall())
    #make sure we got results
    if len(ipString) > 2:
        #results come as a string, convert to a list
        sendList = ipString.split(',')
    # remove extra characters
    for index, item in enumerate(sendList):
        sendList[index] = item.strip(' ()')

    # remove empty items in list
    for i, stuff in enumerate(sendList):
        if stuff == '':
            pass
        else:
            sendList[i] = sendList[i][1:-1]

print "writing send.txt"

#write the order of sending by src IP address
fo = open("send.txt", "w")
send = "hosta\n"
rcv = "hostb\n"
for item in sendList:
```

```
            if item == tgt_sip:
                fo.write(send)
            elif item == tgt_dip:
                fo.write(rcv)
            else:
                pass
        fo.close()  #close the file

##  Disconnect from the DB
    con7.close()
    cursor7.close()


# takes in a list of packets and the host name (hosta or hostb)
# writes a pcap file of all packets that host will send
# saves file as hosta.pcap or hostb.pcap
def writeHost(flowList, hostNo):
    tempList = []
    finalFlowList = []
    index = 0

    #comes as one long list of packets and traits
    # convert to a list of lists where
    # each packet has a list of traits
    for item in flowList:
        if index ==15:
            index = 0
            finalFlowList.append(tempList)
            tempList = []
        tempList.append(item)
        index = index + 1
    finalFlowList.append(tempList)

    #open a file to write pcap to
    hostA_pcap  =  dpkt.pcap.Writer(open('host'  +  hostNo  +  '.pcap',
'wb'))

    cnt = 0
    #write the packets to a file
    for index, item in enumerate(finalFlowList):

        cnt += 1 #increment the count of packets
        #write the packet to the file
        eth_write = writePackets(finalFlowList[index])
        temp = finalFlowList[index][0][10:-2]
        #strip quotes if they are present
        if temp[0] == "'":
            temp = temp[1:]

        tStamp = temp
        # write the file
        hostA_pcap.writepkt(eth_write, Decimal(tStamp))

    # close the file
    hostA_pcap.close()
```

116

```python
def writePackets(pktList):
    # ethernet header
    eth = dpkt.ethernet.Ethernet()
    eth.type = dpkt.ethernet.ETH_TYPE_IP #2048 = IP
    #format the timestamp
    ts = pktList[0][10:-2]
    if ts[0] == "'":
        eth.ts = ts[1:]
    else:
        eth.ts = ts

    pktList[1] = pktList[1].translate(None, ':')
    src = pktList[1]
    src = src[1:13]
    eth.src = binascii.a2b_hex(src)
    pktList[2] = pktList[2].translate(None, ':')
    dst = pktList[2]
    dst = dst[1:13]
    eth.dst = binascii.a2b_hex(dst)

    # IP header
    ip = dpkt.ip.IP()
    srcip = pktList[3]
    srcip = srcip[1:-1]
    dstip = pktList[4]
    dstip = dstip[1:-1]
    ip.dst = socket.inet_aton(dstip)
    ip.src = socket.inet_aton(srcip)
    ip.p = int(pktList[5][:-1])

    # tcp header
    tcp = dpkt.tcp.TCP()
    tcp.flags = int(pktList[7][:-1])
    tcp.sport = int(pktList[8][:-1])
    tcp.dport = int(pktList[9][:-1])
    tcp.seq = long(pktList[10][1:-1])
    tcp.ack = long(pktList[11][1:-1])
    tcp.win = int(pktList[12][:-1])
    data = pktList[13][1:-3]
    tcp.data = binascii.a2b_base64(data)
    topts = pktList[14][1:-4]
    tcp.opts = binascii.a2b_base64(topts)
    tcp.off = (20 + len(tcp.opts)) >> 2


    #perform encapsulation
    ip.data = tcp
    tcp.sum = ip.data.sum
    ip.len = len(ip)
    ip.sum = 0
    eth.data = ip

    return eth

#returns a list of all the packets within a specified TCP flow
```

```
def getTgtFlow(sip, sport, dip, dport):
    global tgtFlowList

    tgtFlowList = []
    print "in getTgtFlow"

    #connect to the database
    con3 = sql.connect("localhost", "root", "Dipper01", "pcapParser")
    cursor3 = con3.cursor()

    # query the DB
    try:
        string1 = "call targetFlow('" + sip + "', " + str(sport) + ", '" + dip + "', " + str(dport) + ")"
        message = cursor3.execute(string1)
        flowString = str(cursor3.fetchall())
        #returns a string, convert to a list and format
        if len(flowString) > 2:
            tgtFlowList = flowString.split(',')

            for index, item in enumerate(tgtFlowList):
                tgtFlowList[index] = item.strip(' ')
            ts = tgtFlowList[0][10:-2]
            if ts[0] == "'":
                ts = ts[1:]
            #perform formatting and save variables to write
            smac = tgtFlowList[1]
            dmac = tgtFlowList[2]
            sip = tgtFlowList[3]
            dip = tgtFlowList[4]
            proto = tgtFlowList[5]
            ipData = tgtFlowList[6]
            flags = tgtFlowList[7]
            sport = tgtFlowList[8]
            dport = tgtFlowList[9]
            seq = tgtFlowList[10]
            ack = tgtFlowList[11]
            rxWin = tgtFlowList[12]
            tcpData = tgtFlowList[13]
            tcpOpts = tgtFlowList[14]
            ts = tgtFlowList[0]
            ts = ts[2:]
            smac = tgtFlowList[1][1:-1]
            dmac = tgtFlowList[2][1:-1]
            sip = tgtFlowList[3][1:-1]
            dip = tgtFlowList[4][1:-1]
            proto = tgtFlowList[5]
            ipData = tgtFlowList[6]
            flags = tgtFlowList[7]
            sport = tgtFlowList[8]
            dport = tgtFlowList[9]
            seq = tgtFlowList[10]
            ack = tgtFlowList[11]
            rxWin = tgtFlowList[12]
            tcpData = tgtFlowList[13]
            tcpOpts = tgtFlowList[14][:-1]
```

```python
        #print results for viewing
##          print "ts:::" + ts
##          print "src mac: " + str(smac)
##          print "dest mac: " + str(dmac)
##          print "srcIP: " +  str(sip)
##          print "dstIP: " + dip
##          print "proto: " + str(proto)
##          print "ipData: " + ipData
##          print "flags: " + str(flags)
##          print "sport: " + str(sport)
##          print "dport: " + str(dport)
##          print "seq: " + str(seq)
##          print "ack: " + str(ack)
##          print "rxWin: " + str(rxWin)
##          print "tcpData: " + tcpData
##          print "tcp Opts: " + tcpOpts


    except Exception as e:
        print "Error: ", e

    cursor3.close()
    con3.close()

    return tgtFlowList

#used to remove empty values from a list
#returned by the DB
def removeVals(theList, val):
    return [value for value in theList if value != val]

#this function takes in a packet ID and
# updates the DB that the packet is a retransmission
def callMarkDupes(pid):
    #open the DB
    con3 =  sql.connect("localhost", "root", "Dipper01", "pcapParser")
    cursor3 = con3.cursor()

    try:
        message = cursor3.execute("call markDup(" + str(pid) + ")")
        con3.autocommit(True)

    except Exception as e:
        print "Error: ", e

    cursor3.close()
    con3.close()

#this finction takes in a TCP sequence number and returns a list of
# all pktIDs that have the same sequence number
def callFindDupes(seq):
    global pktString, pktList

    pktList = []
    indexList = []
```

```python
    #connect to the DB
    con3 =  sql.connect("localhost", "root", "Dipper01", "pcapParser")
    cursor3 = con3.cursor()

    try:  #send the query
        message = cursor3.execute("call noDups(" + str(seq) + ")")
        pktString = str(cursor3.fetchall())
        #results are returned as a string
        #convert to list and format
        if len(pktString) > 2:
            pktList = pktString.split(',')

            for index, item in enumerate(pktList):
                pktList[index] = item.strip(' ()')

            badVal = ''
            # remove empty strings from list of packets
            pktList = removeVals(pktList, badVal)

    except Exception as e:
        print "Error: ", e

    cursor3.close()
    con3.close()

    return pktList

# this function finds a list of unique sequence numbers
def callGetSeqNo():
    global seqString

    seqString = ""

    con3 =  sql.connect("localhost", "root", "Dipper01", "pcapParser")
    cursor3 = con3.cursor()

    try:
        message = cursor3.execute("call getSeqNo()")

        seqString = str(cursor3.fetchall())
        seqList = seqString.split(',')

        for index, item in enumerate(seqList):
            seqList[index] = item.strip(' ()')

        for index, stuff in enumerate(seqList):
            if stuff == '':
                del seqList[index]

    except Exception as e:
        print "Error: ", e

    cursor3.close()
    con3.close()

    return seqList
```

```python
#This function takes in a list of TCP packet traits and enters them
into the TCP table
def callInsertTCP(pktID, flags, isReTx, srcPort, destPort, seqNo,
ackNo, recvWin, tcpData, options):

##      # connect to the database
    con1 = sql.connect("localhost", "root", "Dipper01", "pcapParser")
    cursor1 = con1.cursor()


    if isReTx == False:
        isRetrans = 0
    else:
        isRetrans = 1

    tcpStr = "call insertTCPPackets(" + str(pktID) +  ", " + str(flags)
+ ", " + str(isRetrans) + ", " + str(srcPort) + ", " + str(destPort) +
", " + str(seqNo) + ", " + str(ackNo) + ", "  + str(recvWin) + ", '" +
bytes(tcpData) + "', '" + str(options) +   "')"

    #add packet to database
    try:
        message = cursor1.execute(tcpStr)
        con1.autocommit(True)

        if message == 1:
##            print "added tcp packet to database"
            pass
        else:
##                    print "message: ", message
            print "failed to add packet"
    except Exception as e:
        print "TCP error" + str(pktID)
        print e

    #close the connection
    cursor1.close()
    con1.close()

#this function takes in a list of IP packet traits and writes the to
the IP table
#in the DB
def callInsertIP(packetID, ts, rsrc_mac, rdst_mac, srcIP, dstIP, proto,
ip_data):

##      # connect to the database
    con = sql.connect("localhost", "root", "Dipper01", "pcapParser")
    cursor = con.cursor()
    #used to quickly compare the data
    hashData = hashlib.md5(bytes(ip_data)).hexdigest()

    pktStr = "call insertIPPackets(" + str(packetID) +  ", " + str(ts)
+  ", '" +  rsrc_mac  + "', '"  +  rdst_mac  +  "',  '"  +
socket.inet_ntoa(srcIP) \
```

```
            + "', '" + socket.inet_ntoa(dstIP) + "', " + str(proto) + ", '" +
str(ip_data) + "', '" + bytes(hashData) + "')"

        #add packet to database
        try:
            message = cursor.execute(pktStr)
            con.autocommit(True)

            if message == 1:
##                  print "added ippacket to database"
                pass
            else:
##                              print "message: ", message
                print "failed to add packet"
        except Exception as e:
            print "IP error" + str(packetID)
            print e
            print pktStr

        #close the connection
        cursor.close()
        con.close()

#this function takes in a packet ID
def isRetrans(pid):
    if pid == 1680:
        con2    =    sql.connect("localhost",    "root",    "Dipper01",
"pcapParser")
        cursor2 = con2.cursor()

        try:
            message    =    cursor2.execute("select    *    from
pcapParser.TCPPacketData,    pcapParser.rawPacketData    where
pcapParser.rawPacketData.pid  =  pcapParser.TCPPacketData.pktID   and
pcapParser.rawPacketData.pid = 1680;")
            con2.autocommit(True)
            ipPkt = str(cursor2.fetchall())

            #get ip packet data
            ipList = ipPkt.split(',')

            flag = ipList[1]
            flag = int(flag[1:-1])
            srcPort = ipList[3]
            srcPort = int(srcPort[1:-1])
            dstport = ipList[4]
            dstport = int(dstport[1:-1])
            seq = ipList[5]
            seq = seq[2:-1]
            ack = ipList[6]
            ack = ack[2:-1]
            rxWin = ipList[7]
            rxWin = rxWin[1:-1]
            tcpData = ipList[8]
            tcpData = tcpData[2:-2]
            tcpOpts = ipList[9]
```

```
                    tcpOpts = tcpOpts[2:-3]

                    tstamp = ipList[11]
                    tstamp = float(tstamp[1:])
                    srcMac = ipList[12]
                    srcMac = srcMac[2:-1]
                    dstMac = ipList[13]
                    dstMac = dstMac[2:-1]
                    srcIP = ipList[14]
                    srcIP = srcIP[2:-1]
                    dstIP = ipList[15]
                    dstIP = dstIP[2:-1]
                    proto = ipList[16]
                    proto = int(proto[1:-1])
                    ipData = ipList[17]
                    ipData = ipData[2:-1]

                    fullPacket = [tstamp, srcMac, dstMac, srcIP, dstIP, proto,
ipData,\
                                  flag, srcPort, dstport, seq, ack, rxWin,
tcpData, tcpOpts]
            except Exception as e:
                print e

            cursor2.close()
            con2.close()
            return fullPacket

        else:
            pass




#takes in a binary mac address and converts it to a string
def decode_mac(bin_mac):

    s1,s2,s3,s4,s5,s6 = struct.unpack("BBBBBB",bin_mac)

    if len(hex(s1))<4: s1="0"+str(hex(s1)[2:])
    else: s1=str(hex(s1)[2:])
    if len(hex(s2))<4: s2="0"+str(hex(s2)[2:])
    else: s2=str(hex(s2)[2:])
    if len(hex(s3))<4: s3="0"+str(hex(s3)[2:])
    else: s3=str(hex(s3)[2:])
    if len(hex(s4))<4: s4="0"+str(hex(s4)[2:])
    else: s4=str(hex(s4)[2:])
    if len(hex(s5))<4: s5="0"+str(hex(s5)[2:])
    else: s5=str(hex(s5)[2:])
    if len(hex(s6))<4: s6="0"+str(hex(s6)[2:])
    else: s6=str(hex(s6)[2:])

    d_mac = s1 + ":" + s2 + ":" + s3 + ":" + s4 + ":" + s5 + ":" + s6

    return d_mac
```

```
if __name__ == '__main__':
    main()
```

B.   SOURCE CODE FOR ORDEREDREPLAY.PY

```
'''
Program name: orderedReplay.py
Author:  LT Tom Le Vier
Date: 21 March 2013

Purpose:  This program reads in a text file containing
the ordering of sending and creates a new text file that
contains the host name and how many times it sends. Then it
reads the new text file and if it sees its own hostname then
it will send the appropriate number of times, if it reads hostb
then it will listen the appropriate number of times.  This
program requires send.txt and hostA.pcap to be created from
tcpFLowPrepper.py.  It also requires target_flow.txt to be
configured with the src/dst IP and port numbers and lab_input.txt
tobe configured with src/dst IP and MAC addresses.  You must have
root privileges to run this program.

hostA.pcap contains all the packets this host will transmit.

This program is for Host A.

'''

import dpkt
import dumbnet as dnet
import time
import socket
import pcap
import binascii
import sys
import time

def main():
    global listOfPackets, packetList, cnt, idx, s, p
    global tgt_sip, tgt_sport, tgt_dip, tgt_dport

    # look for packets from this flow
    tgtFile = open("target_flow.txt", "r")
    tgt_sip = tgtFile.readline()
    tgt_sip = tgt_sip[:-1]
    tgt_sport = tgtFile.readline()
    tgt_sport = tgt_sport[:-1]
    tgt_dip = tgtFile.readline()
    tgt_dip = tgt_dip[:-1]
    tgt_dport = tgtFile.readline()
    tgt_dport = tgt_dport[:-1]
```

124

```
#create pcap object to listen
prt_filter = "dst port " + str(tgt_dport)

p = pcap.pcapObject()
dev = pcap.lookupdev()
p.open_live(dev, 1600, 0, 100)
p.setfilter(prt_filter, 0, 0)

#create the socket
try:
    s = dnet.ip()
except socket.error, msg:
    print "Failed to create socket. Error: " + str(msg[0]) +
"\nError Message: " + msg[1]
    sys.exit()

#write a file with the order of sending and receiving
writeOrder()

#when packet has been received
rxFlag = True

#only parse up to this many pkts
# if -1, will parse entire file
maxPktsToProc = -1


##read in ip/mac addresses for lab setup
print "\nlab Set up: "
labfile = open("lab_input.txt", "r")
lab_sip = labfile.readline()
lab_sip = lab_sip[:-1]
lab_smac = labfile.readline()
lab_smac = lab_smac[:-1]
lab_dip = labfile.readline()
lab_dip = lab_dip[:-1]
lab_dmac = labfile.readline()
lab_dmac = lab_dmac[:-1]
lab_smac = lab_smac.translate(None, ':')
lab_dmac = lab_dmac.translate(None, ':')
print lab_smac

#opensend file to see who's turn it is to send
orderFile = open("order.txt", "r")

#trim off magical 13th character from reading in
lab_smac = lab_smac[0:12]
lab_dmac = lab_dmac[0:12]
print "lab_srcIP =", lab_sip
print "length ", len(lab_sip)
print "lab src_pmac =", lab_smac
print "lab dst_ip =", lab_dip
print "lab dst_mac =", lab_dmac

fileName = "hostA.pcap"
```

```
#open capture file to read
pcr = dpkt.pcap.Reader(open(fileName))

listOfPackets = []

##   read in packet data using timestamp and buffer
for ts, buf in pcr:
    packetList = []
    #read ethernet data from the buffer
    eth_in = dpkt.ethernet.Ethernet(buf)
    ip_in = dpkt.ip.IP()
    ip_in = eth_in.data
    tcp_in = dpkt.tcp.TCP()
    tcp_in = ip_in.data

    # add packet chars to packetList
    packetList.append(binascii.a2b_hex(lab_smac)) #src mac 0
    packetList.append(binascii.a2b_hex(lab_dmac)) # dst mac 1
    packetList.append(socket.inet_aton(lab_sip)) # src IP 2
    packetList.append(socket.inet_aton(lab_dip)) # dst IP 3
    packetList.append(ip_in.ttl)            # time to live 4
    packetList.append(ip_in.opts)           # IP options 5
    packetList.append(ip_in.tos)            # type of service 6
    packetList.append(ip_in.p)              # protocol 7
##      packetList.append(dpkt.tcp.TCP())            # ip.data = TCP
    packetList.append(tcp_in.flags)         # tcp flags 8
    packetList.append(tcp_in.sport)         # src port 9
    packetList.append(tcp_in.dport)         # dst port 10
    packetList.append(tcp_in.seq)           # sequence no 11
    packetList.append(tcp_in.ack)           # ack no 12
    packetList.append(tcp_in.win)           # receiver win size 13
    packetList.append(tcp_in.data)          # tcp data 14
    packetList.append(tcp_in.opts)          # tcp options 15
    packetList.append(tcp_in.sum)           # check sum 16
    packetList.append(tcp_in.urp)           # urgent pointer 17
    packetList.append(tcp_in.off)           # tcp offset 18


    #add packet to listOf Packets
    listOfPackets.append(packetList)

start = 0  #timer to check for deadlock  #not currently implemented
elapsed = 0  #timer to check for deadlock

while(True):
    #check flag
    if rxFlag == True:
        start = 0  #reset timer

        #read in a line
        tx = orderFile.readline()[:-1]

        #check if the file is empty, if so close
        if not tx:
            print "all out of packets"
```

```
                orderFile.close()
                break

            #save host and number of times to transmit or receive
            host = tx[:5]
            cnt = int(tx[6:])
            setCnt(cnt)

            #if hosta -->transmit a packet
            if host == "hosta":
                index = 0
                while index < cnt:
                    packet = listOfPackets.pop(0)
                    pause()
                    print "sending..."
                    sendPacket(packet)
                    index = index + 1 #increment index to send right
number

            #if host B listen for packets
            elif host == "hostb":
                start = time.time()  #start the timer
                print "listening 1"
                idx = 0
                rxFlag = listen(cnt)


def sendPacket(list1):
    global s
    #Instatiate IP and TCP
    ip_out = dpkt.ip.IP()
    tcp_out = dpkt.tcp.TCP()

    #build the tcp portion of the packet
    tcp_out.off = list1[18]
    tcp_out.urp = list1[17]
    tcp_out.opts = list1[15]
    tcp_out.data = list1[14]
    tcp_out.win = list1[13]
    tcp_out.ack = list1[12]
    tcp_out.seq = list1[11]
    tcp_out.dport = list1[10]
    tcp_out.sport = list1[9]
    tcp_out.flags = list1[8]


    #build the ip portion of the packet
    ip_out.data = tcp_out
    ip_out.src = list1[2]
    ip_out.dst = list1[3]
    ip_out.ttl = list1[4]
    ip_out.opts = list1[5]
    ip_out.tos = list1[6]
    ip_out.p = list1[7]
    ip_out.len = len(ip_out)
    buf = dnet.ip_checksum(str(ip_out))
```

```
        #print some stuff for ts
        print 'sport: ', tcp_out.sport, '\tdport: ', tcp_out.dport
        print 'seq', tcp_out.seq, '\nack', tcp_out.ack

        #send the packet
        s.send(buf)

#create a delay
def pause():
    start = time.time()
    elapsed = 0

    while elapsed < .8:
        elapsed = time.time() - start


#bring in binary ethrnet address and convert it to a string
def decode_mac(bin_mac):

    s1,s2,s3,s4,s5,s6 = struct.unpack("BBBBBB",bin_mac)

    if len(hex(s1))<4: s1="0"+str(hex(s1)[2:])
    else: s1=str(hex(s1)[2:])
    if len(hex(s2))<4: s2="0"+str(hex(s2)[2:])
    else: s2=str(hex(s2)[2:])
    if len(hex(s3))<4: s3="0"+str(hex(s3)[2:])
    else: s3=str(hex(s3)[2:])
    if len(hex(s4))<4: s4="0"+str(hex(s4)[2:])
    else: s4=str(hex(s4)[2:])
    if len(hex(s5))<4: s5="0"+str(hex(s5)[2:])
    else: s5=str(hex(s5)[2:])
    if len(hex(s6))<4: s6="0"+str(hex(s6)[2:])
    else: s6=str(hex(s6)[2:])

    d_mac = s1 + ":" + s2 + ":" + s3 + ":" + s4 + ":" + s5 + ":" + s6

    return d_mac

#listens for packets from the other host
def listen(cnt):
    global idx

    #set the flag
    dis = False

    #listen for the right number of packets
    while(idx < cnt):
        try:
            #check to see if it is an Ethernet packet
            if p.datalink() == pcap.DLT_EN10MB: #
                try:
                    #process the Ethernet packet
                    dis = p.dispatch(1, processOneEthPkt)
                except Exception as e:
                    print "error", e
```

```
            except KeyboardInterrupt:
                print '%s' % sys.exc_type
                print 'shutting down'
                sys.exit(0)

            except Exception as e:
                print e

    return True

def processOneEthPkt(pktlen, data, timestamp):
    global idx

    print "processing..."

    #If there is no Ethernet data, wrong packet
    if not data:
        print "no data"
        return False
    else:
        eth = dpkt.ethernet.Ethernet(data)
        #print some out output
        print "received packet from "
        print "src IP: ", socket.inet_ntoa(eth.data.src)
        print "dest port: ", eth.data.data.sport, "\nseq no: ",
eth.data.data.seq
        idx = idx + 1 #keep track of how many packets received
        print idx

#setters and getters
def setCnt(count):
    global cnt
    cnt = count

def getCnt():
    return cnt

# this function reads in a textfile containing the
# order of transmission and outputs another text
# file that contains the order and the number of
# times for that host to send
# hosta 1 --> send 1 packet
# hostb 2--> listen for 2 packets
def writeOrder():
    ##  initialize vars
    temp = ''
    ctr = 1

    #opensend file to see who's turn it is to send
    sendFile = open("send.txt", "r")
    #open file to write to
    orderFile = open("order.txt", "w")

    host = sendFile.readline()[:-1]
    temp = host
```

```
        while(True):
            host = sendFile.readline()[:-1]
            #if send.txt has no more lines to read
            if not host:
                string1 = temp + " " + str(ctr) + "\n"
                orderFile.write(string1)
                sendFile.close()
                orderFile.close()
                break

            if host == temp:
                ctr = ctr + 1

            else:
                string1 = temp + " " + str(ctr) + "\n"
                orderFile.write(string1)
                temp = host
                ctr = 1


if __name__ == '__main__':
    main()
```

C.   SOURCE CODE FOR FLOWSTATS.PY

```
##   Authors:  Tom LeVier
##   Date:  23 March 2012
##   File Name: flowStats.py
##   Masters Thesis
##   "A Tool for Stateful TCP Replay"
##   Thesis advisors:
##     Professor Xie
##     Professor Gibson
##
##   Purpose:  The purpose of this program is to
##   read in a .pcap file and give statistics.
##
##   Uses Python 2.7 on ubuntu 11.10


import sys
import dpkt
import pcap
import socket
import binascii
import struct
import MySQLdb as sql
import array
import hashlib
from collections import defaultdict

#track number of packets from src port
counter = 0
```

```python
#port number to extract flow from
target_portNo = 0


##  This splits up an IP address and converts it from in to char
##  so it is readable
def ipv4addr(addr):
    addrcode = [chr(int(i)) for i in addr.split('.')]
    return "".join(addrcode)

def main():
    global counter, src_ip, dst_ip, src_mac, dst_mac

    #pcap file to read in
    filename = "pcap6_2.pcap"

    #output file to write to
    fo = open("test4.txt", "w")

    #variables
    notIPcount = 0 # track number of packets that are not IP
    tcpCount = 0   #track tcp packets
    ipCount = 0    #track IP packets
    totalSize = 0.0  #track total size of packet data
    avgSize = 0.0  # avg packet size
    flowList = []
    flowID = {}
    flowDict = defaultdict(list)  # dictionary of unique tcp flows
    dupDict = defaultdict(list)  #dictionary of duplicate tcp flows
    hostApkts = 0 #packets sent by host A
    hostBpkts = 0 #packets sent by host B
    flowTotalSize = 0.0  #total amount of data for tcp flow
    flowAverageSize = 0.0  #average packet size for tcp flow
    flowCnt = 1      # number of unique tcp flows
    numFlows = 0.0


    # read in the target TCP parameters
    tgtfile = open("target_flow.txt", "r")
    tgt_sip = tgtfile.readline()
    tgt_sip = tgt_sip[:-1]
    tgt_sport = tgtfile.readline()
    tgt_sport = tgt_sport[:-1]
    tgt_dip = tgtfile.readline()
    tgt_dip = tgt_dip[:-1]
    tgt_dport = tgtfile.readline()
    tgt_dport = tgt_dport[:-1]


    #open capture file to read
    pcr = dpkt.pcap.Reader(open(filename))

    #initialize packetID
    packetID = 1
```

131

```
    ##  read in packet data using timestamp and buffer
    for ts, buf in pcr:
        #read ethernet data from the buffer
        eth_in = dpkt.ethernet.Ethernet(buf)

        #check mac addresses from input file, don't do anything with
them
        rdst_mac = decode_mac(eth_in.dst)
        rsrc_mac = decode_mac(eth_in.src)


        # determine if packet has IP data
        if eth_in.type == 2048:
            ipCount += 1

            #track avg packet size
            totalSize = totalSize + len(eth_in.data)


            #save IP data
            ip_in = eth_in.data


            #if IP data has TCP
            if ip_in.p == dpkt.ip.IP_PROTO_TCP:
                #increment tcpCount
                tcpCount += 1

                #convert IP address to string
                srcip = socket.inet_ntoa(ip_in.src)
                dstip = socket.inet_ntoa(ip_in.dst)

                #encapsulate TCP in IP
                tcp = ip_in.data

                dport = tcp.dport
                sport = tcp.dport

                #tuple for tcp flow src IP, dst IP, src port, dst port
                tup = srcip + dstip + str(sport) + str(dport)
                dup = dstip + srcip + str(dport) + str(sport)

                #if flow is not in flow dict and not in dup dict
                #then add the flow
                if (tup not in flowDict) and (dup not in dupDict):
                    flowDict[tup].append
                    dupDict[dup].append


                # match the target flow
                if (tgt_sip == srcip or tgt_sip == dstip):
                    if (tgt_dip == srcip or tgt_dip == dstip):
                        if    (int(tgt_sport)    ==    tcp.sport    or
int(tgt_sport) == tcp.dport):
```

```python
                           if   (int(tgt_dport)   ==   tcp.sport   or
int(tgt_dport) == tcp.dport):

                                    # increment # of packets in flow
                                    flowCnt = flowCnt + 1

                                    #track total size of data
                                    flowTotalSize    =    flowTotalSize    +
len(eth_in.data)

                                    #increment packet count for each host
                                    if srcip == tgt_sip:
                                        hostApkts = hostApkts + 1
                                    elif srcip == tgt_dip:
                                        hostBpkts = hostBpkts + 1

        else:
            #not ip packet
            notIPcount = notIPcount + 1 #tracknumber of packets that
are not IP

    #determine packet statistics
    avgSize = totalSize / ipCount
    flowAvgSize = flowTotalSize / flowCnt

    #  print some results
    fo.write( "From " + filename + "\n")
    fo.write("******************\nResults            for            entire
trace\n******************"+ "\n")
    fo.write( "\nFound "+ str(ipCount)+ " that were IP packets"+ "\n")
    fo.write( "Found " + str(notIPcount) + " that were not IP packets"+
"\n")
    fo.write("Found " + str(tcpCount) + " that were TCP packets"+ "\n")
    fo.write("avgSize = totalSize / ipCount\n")
    fo.write("total = " + str(totalSize)+ "\n")
    fo.write("Average size of packet for entire trace: " + str(avgSize)
+ "bytes"+ "\n")
    fo.write("found" + str(len(flowDict)) + "unique TCP Flows"+ "\n")
    fo.write("\n******************\nResults      for      target      TCP
Flow\n******************\n")
    fo.write("tgt_srcIP =" + tgt_sip+ "\n")
    fo.write("tgt_src_port = " + str(tgt_sport)+ "\n")
    fo.write("dst_ip =" + tgt_dip+ "\n" )
    fo.write("tgt_dst_port = " + str(tgt_dport) + "\n")
    fo.write( "Target flow contained" + str( flowCnt - 1) + "packets"+
"\n")
    fo.write(tgt_sip + " transmitted" + str(hostApkts) + "packets"+
"\n")
    fo.write(tgt_dip + " transmitted" + str(hostBpkts) + "packets"+
"\n")
    fo.write("flowAvgSize = flowTotalSize / flowCnt\n")
    fo.write("Average  size  of  packet  for  Target  TCP  flow:  "  +
str(flowAvgSize) + "bytes"+ "\n")
    fo.write("total size = " + str(flowTotalSize)+ "\n")

    fo.close()
```

133

```python
        print "complete"




def removeVals(theList, val):
    return [value for value in theList if value != val]




def decode_mac(bin_mac):

    s1,s2,s3,s4,s5,s6 = struct.unpack("BBBBBB",bin_mac)

    if len(hex(s1))<4: s1="0"+str(hex(s1)[2:])
    else: s1=str(hex(s1)[2:])
    if len(hex(s2))<4: s2="0"+str(hex(s2)[2:])
    else: s2=str(hex(s2)[2:])
    if len(hex(s3))<4: s3="0"+str(hex(s3)[2:])
    else: s3=str(hex(s3)[2:])
    if len(hex(s4))<4: s4="0"+str(hex(s4)[2:])
    else: s4=str(hex(s4)[2:])
    if len(hex(s5))<4: s5="0"+str(hex(s5)[2:])
    else: s5=str(hex(s5)[2:])
    if len(hex(s6))<4: s6="0"+str(hex(s6)[2:])
    else: s6=str(hex(s6)[2:])

    d_mac = s1 + ":" + s2 + ":" + s3 + ":" + s4 + ":" + s5 + ":" + s6

    return d_mac


if __name__ == '__main__':
    main()
```

LIST OF REFERENCES

[1] Y. Lin, P. Lin, T. Cheng, I. Chen and Y. Lai, "Low-storage capture and loss recovery selective replay of real flows," *IEEE Communications Magazine,* vol. 50, no. 4, pp. 114-121, 5 April 2012.

[2] Y. Cheng, U. Holzle, S. Savage and G. M. Voelker, "Monkey see, monkey do: A tool for TCP tracing and replaying," *Usenix 2004 Annual Technical Conference*, pp 87-98, June 2004.

[3] IXIA, "IXIA Breaking-Point Storm," (2013). [Online]. Available http://www.ixiacom.com/products/network_test/breakingpoint/product-line/breakingpoint-storm/index.php

[4] S. Jensen, "Consolidated Tactical Network Analysis for Optimizing Bandwidth: Marine Corps Support Wide Area Network (SWAN) and TCP Accelerators," M.S. Thesis, Dept. Inf. Sci., NPS, Monterey, CA, 2009.

[5] J. Postel. (1981, September). "Transmission Control Protocol. RFC 793," [Online]. Available: http://tools.ietf.org/pdf/rfc793.pdf.

[6] G. R. Wright and W. R. Stevens, *TCP/IP Illustrated*, vol. 2, New York: Addison-Wesley, 1995, pp. 98-102.

[7] V. Paxson and M. Allman. (2000, November). "Computing TCP's Retransmission Timer. RFC 2988," [Online]. Available: https://tools.ietf.org/html/rfc2988.

[8] J. F. Kurose and K. Ross, *Computer Networking A Top-Down Approach*, fifth ed., New York: Addison-Wesley, 2010, pp. 197-290.

[9] M. Allman, V. Paxson and E. Blanton. (2009, September). "TCP Congestion Control. RFC 5681," [Online]. Available: http://tools.ietf.org/html/rfc5681.

[10] J. Postel. (1983, November). "TCP Maximum Segment Size and Related Topics. RFC 879," [Online]. Available: http://tools.ietf.org/html/rfc879.

[11] V. Jacobsen, R. Braden and D. Borman. (1992, May). "TCP Extensions for High Performance. RFC 1323," [Online]. Available: http://tools.ietf.org/pdf/rfc1323.pdf.

[12] S. M. Bellovin, "Security problems in the TCP/IP protocol suite," *ACM Computer Communications Review,* Vol. 19, No. 2, pp. 32-48, April 1989.

[13] V. Jacobsen and R. Braden. (1988, October). "TCP Extensions for Long Delay Paths. RFC 1072," [Online]. Available: http://tools.ietf.org/pdf/rfc1072.pdf.

[14] M. Mathis, J. Mahdavi, S. Floyd and A. Romanow. (1996, October). "TCP Selective Acknowledgement Options. RFC 2018," [Online]. Available: http://tools.ietf.org/html/rfc2933.

[15] C. Zeeh. (2003, January). "The Lempel Ziv Algorithm," [Online]. Available: http://tuxtina.de/files/seminar/LempelZiv.pdf.

[16] P. J. Young. (2012, Apr. 25). *MCTSSA Stateful Replay of Captured Data*[Online]. Available e-mail: xie@nps.edu Message: Stateful Replay Requirements.

[17] D. Song. (2006, October). "Package DPKT," [Online]. Available: http://www.monkey.org/~dugsong/dpkt/pydoc/public/dpkt-module.html.

[18] D. Song. (2006, November). "tcp.py," [Online]. Available: http://code.google.com/p/dpkt/source/browse/trunk/dpkt/tcp.py?r=42.

[19] D. Song. (2007, March). "ip.py," [Online]. Available: http://code.google.com/p/dpkt/source/browse/trunk/dpkt/ip.py?r=53.

[20] Pylibpcap. "Pylibpcap: Python Module for Pcap," (2013). [Online]. Available: http://pylibpcap.sourceforge.net/.